Geometry: A Comprehensive Course

Geometry: A Comprehensive Course

Diana Marks

Larsen & Keller
www.larsen-keller.com

Geometry: A Comprehensive Course
Diana Marks
ISBN: 978-1-64172-681-8 (Hardback)

⊟ Larsen & Keller

Published by Larsen and Keller Education,
5 Penn Plaza,
19th Floor,
New York, NY 10001, USA

Cataloging-in-Publication Data

Geometry : a comprehensive course / Diana Marks.
      p. cm.
Includes bibliographical references and index.
ISBN 978-1-64172-681-8
1. Geometry. 2. Mathematics. 3. Euclid's Elements. I. Marks, Diana.
QA445 .G46 2022
516--dc23

For more information regarding Larsen and Keller Education and its products, please visit the publisher's website www.larsen-keller.com

Table of Contents

Preface

This book has been written, keeping in view that students want more practical information. Thus, my aim has been to make it as comprehensive as possible for the readers. I would like to extend my thanks to my family and co-workers for their knowledge, support and encouragement all along.

Geometry is a mathematical branch that deals with the questions of size, shape, relative position of figures and the properties of space. Some of the fundamental concepts of geometry include concepts of line, surface, point, plane, angle and curve as well as the notions of manifold and topology. Geometry is applied in many fields such as architecture, physics, art and other branches of mathematics. A few of the sub-disciplines within this field are Euclidean geometry, differential geometry and algebraic geometry. Euclidean geometry is applied in the fields of computer graphics, computational geometry, incidence, geodesy and navigation. Differential geometry involves the usage of techniques of calculus and linear algebra for studying geometric problems. The topics included in this book on geometry are of utmost significance and bound to provide incredible insights to readers. It elucidates new techniques and their applications in a multidisciplinary approach. Those in search of information to further their knowledge will be greatly assisted by this book.

A brief description of the chapters is provided below for further understanding:

Chapter – Introduction

Geometry is a branch of mathematics which deals with the study of points, lines, surfaces, shapes, size, etc. Algebraic geometry, analytical geometry, finite geometry, complex geometry, convex geometry, digital geometry, etc. are some of the branches of geometry. All the diverse branches of geometry have been briefly analyzed in this chapter.

Chapter – Fundamental Concepts of Geometry

Some of the basic elements of geometry are point, lines, acute angle, obtuse angle, right angle, whole angle, etc. The topics elaborated in this chapter will help in gaining a better perspective about these elements of geometry.

Chapter – Geometrical Figures

There are a number of figures in geometry that are essential in the concepts and principles of mathematics. Scalene Triangle, Isosceles Triangle, Equilateral Triangle, Right Triangle, Square, Rhombus, Kite, Parallelogram, Trapezium, Sphere, Cone, Cuboid, etc. This chapter closely examines these figures of geometry to provide an extensive understanding of the subject.

Chapter – Theorems in Geometry

A hypothesis which is assumed and proved by a chain of reasoning is called a theorem. Geometry consists of several theorems such as Ptolemy's theorem, Brahmagupta's formula, Cauchy's theorem, Finsler–Hadwiger theorem, etc. This chapter has been carefully written to provide an easy understanding of these various theorems of geometry.

Chapter – Conic Section

Conic section is a shape which is formed by the intersection of a cone and a plane. A conic section can attain different shapes depending upon the angle of intersection such as of a circle, an ellipse, a parabola or a hyperbola. These diverse conic sections have been thoroughly discussed in this chapter.

Chapter – Euclidean Geometry

Euclidean geometry refers to the study and analysis of solid shapes, figures and planes, on the basis of axioms, postulates and theorems given by Greek mathematician Euclid. This chapter delves into detailed study of the axioms of Euclidean plane geometry, Euclid's postulates, etc. to provide in-depth understanding of Euclidean geometry.

Diana Marks

1

Introduction

Geometry is a branch of mathematics which deals with the study of points, lines, surfaces, shapes, size, etc. Algebraic geometry, analytical geometry, finite geometry, complex geometry, convex geometry, digital geometry, etc. are some of the branches of geometry. All the diverse branches of geometry have been briefly analyzed in this chapter.

GEOMETRY

Geometry is an original field of mathematics, and is indeed the oldest of all sciences, going back at least to the times of Euclid, Pythagoras, and other "natural philosophers" of ancient Greece. Initially, geometry was studied to understand the physical world we live in, and the tradition continues to this day. Witness for example, the spectacular success of Einstein's theory of general relativity, a purely geometric theory that describes gravitation in terms of the curvature of a four-dimensional "spacetime". However, geometry transcends far beyond physical applications, and it is not unreasonable to say that geometric ideas and methods have always permeated every field of mathematics.

In modern language, the central object of study in geometry is a manifold, which is an object that may have a complicated overall shape, but such that on small scales it "looks like" ordinary space of a certain dimension. For example, a 1-dimensional manifold is an object such that small pieces of it look like a line, although in general it looks like a curve rather than a straight line. A 2-dimensional manifold, on small scales, looks like a (curved) piece of paper – there are two independent directions in which we can move at any point. For example, the surface of the Earth is a 2-dimensional manifold. An n-dimensional manifold likewise looks locally like an ordinary n-dimensional space. This does not necessarily correspond to any notion of "physical space". As an example, the data of the position and velocity of N particles in a room is described by 6N independent variables, because each particle needs 3 numbers to describe its position and 3 more numbers to describe its velocity. Hence, the "configuration space" of this system is a 6N-dimensional manifold. If for some reason the motion of these particles were not independent but rather constrained in some way, then the configuration space would be a manifold of smaller dimension.

Usually, the set of solutions of a system of partial differential equations has the structure of some high dimensional manifold. Understanding the "geometry" of this manifold often gives new insight into the nature of these solutions, and to the actual phenomenon that is modeled by the differential equations, whether it comes from physics, economics, engineering, or any other quantitative science.

A typical problem in geometry is to "classify" all manifolds of a certain type. That is, we first decide which kinds of manifolds we are interested in, then decide when two such manifolds should basically be considered to be the same, or "equivalent", and finally try to determine how many inequivalent types of such manifolds exist. For example, we might be interested in studying surfaces (2-dimensional manifolds) that lie inside the usual 3-dimensional space that we can see, and we might decide that two such surfaces are equivalent if one can be "transformed" into the other by translations or rotations. This is the study of the Riemannian geometry of surfaces immersed in 3-space, and was classically the first subfield of "differential geometry", pioneered by mathematical giants such as Gauss and Riemann in the 1800's.

DIFFERENTIAL GEOMETRY

Differential geometry is the branch of mathematics that studies the geometry of curves, surfaces, and manifolds (the higher-dimensional analogs of surfaces). The discipline owes its name to its use of ideas and techniques from differential calculus, though the modern subject often uses algebraic and purely geometric techniques instead. Although basic definitions, notations, and analytic descriptions vary widely, the following geometric questions prevail: How does one measure the curvature of a curve within a surface (intrinsic) versus within the encompassing space (extrinsic)? How can the curvature of a surface be measured? What is the shortest path within a surface between two points on the surface? How is the shortest path on a surface related to the concept of a straight line?

While curves had been studied since antiquity, the discovery of calculus in the 17th century opened up the study of more complicated plane curves—such as those produced by the French mathematician René Descartes with his "compass". In particular, integral calculus led to general solutions of the ancient problems of finding the arc length of plane curves and the area of plane figures. This in turn opened the stage to the investigation of curves and surfaces in space—an investigation that was the start of differential geometry.

Some of the fundamental ideas of differential geometry can be illustrated by the strake, a spiraling strip often designed by engineers to give structural support to large metal cylinders such as smokestacks. A strake can be formed by cutting an annular strip (the region between two concentric circles) from a flat sheet of steel and then bending it into a helix that spirals around the cylinder, as illustrated in the figure. What should the radius r of the annulus be to produce the best fit? Differential geometry supplies the solution to this problem by defining a precise measurement for the curvature of a curve; then r can be adjusted until the curvature of the inside edge of the annulus matches the curvature of the helix.

An important question remains: Can the annular strip be bent, without stretching, so that it forms a strake around the cylinder? In particular, this means that distances measured along the surface (intrinsic) are unchanged. Two surfaces are said to be isometric if one can be bent (or transformed) into the other without changing intrinsic distances. (For example, because a sheet of paper can be rolled into a tube without stretching, the sheet and tube are "locally" isometric—only locally because new, and possibly shorter, routes are created by connecting the two edges of the paper). Thus, the second question becomes: Are the annular strip and the strake isometric? To answer this and similar questions, differential geometry developed the notion of the curvature of a surface.

Curvature of Curves

Although mathematicians from antiquity had described some curves as curving more than others and straight lines as not curving at all, it was the German mathematician Gottfried Leibniz who, in 1686, first defined the curvature of a curve at each point in terms of the circle that best approximates the curve at that point. Leibniz named his approximating circle the osculating circle. He then defined the curvature of the curve (and the circle) as 1/r, where r is the radius of the osculating circle. As a curve becomes straighter, a circle with a larger radius must be used to approximate it, and so the resulting curvature decreases. In the limit, a straight line is said to be equivalent to a circle of infinite radius and its curvature defined as zero everywhere. The only curves in ordinary Euclidean space with constant curvature are straight lines, circles, and helices. In practice, curvature is found with a formula that gives the rate of change, or derivative, of the tangent to the curve as one moves along the curve. This formula was discovered by Isaac Newton and Leibniz for plane curves in the 17th century and by the Swiss mathematician Leonhard Euler for curves in space in the 18th century. Note that the derivative of the tangent to the curve is not the same as the second derivative studied in calculus, which is the rate of change of the tangent to the curve as one moves along the x-axis.

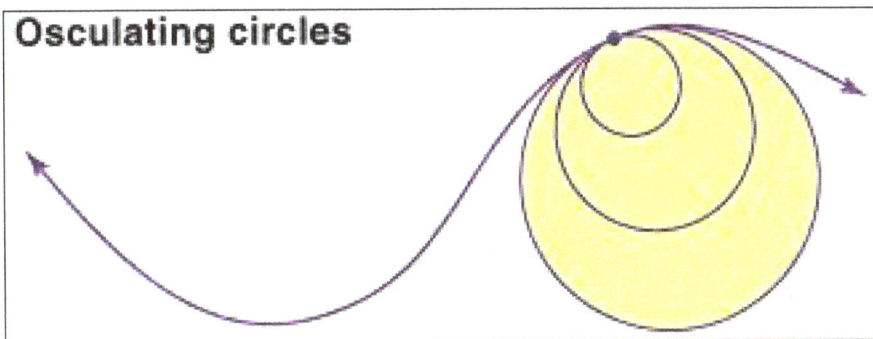

The curvature at each point of a line is defined to be 1/r, where r is the radius of the osculating, or "kissing," circle that best approximates the line at the given point.

With these definitions in place, it is now possible to compute the ideal inner radius r of the annular strip that goes into making the strake shown in the figure. The annular strip's inner curvature 1/r must equal the curvature of the helix on the cylinder. If R is the radius of the cylinder and H is the height of one turn of the helix, then the curvature of the helix is $4\pi^2R/[H^2 + (2\pi R)^2]$. For example, if R = 1 metre and H = 10 metres, then r = 3.533 metres.

Curvature of Surfaces

To measure the curvature of a surface at a point, Euler, in 1760, looked at cross sections of the surface made by planes that contain the line perpendicular (or "normal") to the surface at the point. Euler called the curvatures of these cross sections the normal curvatures of the surface at the point. For example, on a right cylinder of radius r, the vertical cross sections are straight lines and thus have zero curvature; the horizontal cross sections are circles, which have curvature 1/r. The normal curvatures at a point on a surface are generally different in different directions. The maximum and minimum normal curvatures at a point on a surface are called the principal (normal) curvatures, and the directions in which these normal curvatures occur are called the principal directions. Euler proved that for most surfaces where the normal curvatures are not constant (for example, the cylinder), these principal directions are perpendicular to each other. Note that on a sphere all the normal curvatures are the same and thus all are principal curvatures. These principal normal curvatures are a measure of how "curvy" the surface is.

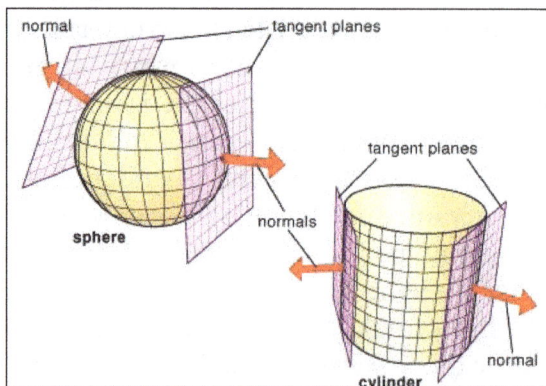

The normal, or perpendicular, at each point of a surface defines the corresponding tangent plane, and vice versa.

The theory of surfaces and principal normal curvatures was extensively developed by French geometers led by Gaspard Monge. It was in an 1827 paper, however, that the German mathematician Carl Friedrich Gauss made the big breakthrough that allowed differential geometry to answer the question raised above of whether the annular strip is isometric to the strake. The Gaussian curvature of a surface at a point is defined as the product of the two principal normal curvatures; it is said to be positive if the principal normal curvatures curve in the same direction and negative if they curve in opposite directions. Normal curvatures for a plane surface are all zero, and thus the Gaussian curvature of a plane is zero. For a cylinder of radius r, the minimum normal curvature is zero (along the vertical straight lines), and the maximum is 1/r (along the horizontal circles). Thus, the Gaussian curvature of a cylinder is also zero.

If the cylinder is cut along one of the vertical straight lines, the resulting surface can be flattened (without stretching) onto a rectangle. In differential geometry, it is said that the plane and cylinder are locally isometric. These are special cases of two important theorems:

- Gauss's "Remarkable Theorem". If two smooth surfaces are isometric, then the two surfaces have the same Gaussian curvature at corresponding points. (Athough defined extrinsically, Gaussian curvature is an intrinsic notion).

- Minding's theorem. Two smooth ("cornerless") surfaces with the same constant Gaussian curvature are locally isometric.

As corollaries to these theorems:

- A surface with constant positive Gaussian curvature c has locally the same intrinsic geometry as a sphere of radius Square root of $\sqrt{1/c}$. (This is because a sphere of radius r has Gaussian curvature $1/r^2$).

- A surface with constant zero Gaussian curvature has locally the same intrinsic geometry as a plane. (Such surfaces are called developable).

- A surface with constant negative Gaussian curvature c has locally the same intrinsic geometry as a hyperbolic plane.

The Gaussian curvature of an annular strip (being in the plane) is constantly zero. So to answer whether or not the annular strip is isometric to the strake, one needs only to check whether a strake has constant zero Gaussian curvature. The Gaussian curvature of a strake is actually negative, hence the annular strip must be stretched—although this can be minimized by narrowing the shapes.

Shortest Paths on a Surface

From an outside, or extrinsic, perspective, no curve on a sphere is straight. Nevertheless, the great circles are intrinsically straight—an ant crawling along a great circle does not turn or curve with respect to the surface. About 1830 the Estonian mathematician Ferdinand Minding defined a curve on a surface to be a geodesic if it is intrinsically straight—that is, if there is no identifiable curvature from within the surface. A major task of differential geometry is to determine the geodesics on a surface. The great circles are the geodesics on a sphere.

A great circle arc that is longer than a half circle is intrinsically straight on the sphere, but it is not the shortest distance between its endpoints. On the other hand, the shortest path in a surface is not always straight, as shown in the figure. An important theorem is:

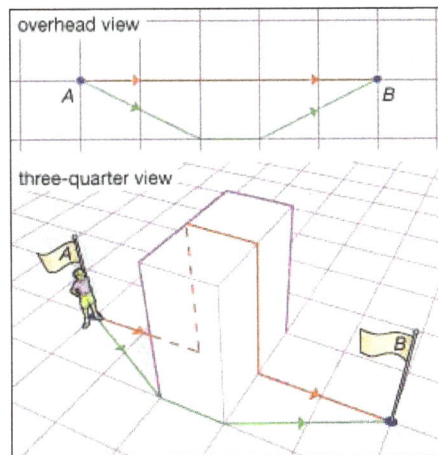

The shaded elevation and the surrounding plane form one continuous surface. Therefore, the red path from A to B that rises over the elevation is intrinsically straight (as viewed from within the surface). However, it is longer than the intrinsically bent green path, demonstrating that an intrinsically straight line is not necessarily the shortest distance between two points.

On a surface which is complete (every geodesic can be extended indefinitely) and smooth, every shortest curve is intrinsically straight and every intrinsically straight curve is the shortest curve between nearby points.

CONVEX GEOMETRY

In mathematics, convex geometry is the branch of geometry studying convex sets, mainly in Euclidean space. Convex sets occur naturally in many areas: computational geometry, convex analysis, discrete geometry, functional analysis, geometry of numbers, integral geometry, linear programming, probability theory, game theory, etc.

Classification

According to the Mathematics Subject Classification MSC2010, the mathematical discipline *Convex and Discrete Geometry* includes three major branches:

- General convexity.

- Polytopes and polyhedra.

- Discrete geometry.

(though only portions of the latter two are included in convex geometry).

General convexity is further subdivided as follows:

- Axiomatic and generalized convexity.

- Convex sets without dimension restrictions.

- Convex sets in topological vector spaces.

- Convex sets in 2 dimensions (including convex curves).

- Convex sets in 3 dimensions (including convex surfaces).

- Convex sets in n dimensions (including convex hypersurfaces).

- Finite-dimensional Banach spaces.

- Random convex sets and integral geometry.

- Asymptotic theory of convex bodies.

- Approximation by convex sets.

- Variants of convex sets (star-shaped, (m, n)-convex, etc.)

- Helly-type theorems and geometric transversal theory.

- Other problems of combinatorial convexity.

- Length, area, volume.

- Mixed volumes and related topics.

- Inequalities and extremum problems.

- Convex functions and convex programs.

- Spherical and hyperbolic convexity.

The term *convex geometry* is also used in combinatorics as an alternate name for an antimatroid, which is one of the abstract models of convex sets.

ALGEBRAIC GEOMETRY

Algebraic geometry is the study of the geometric properties of solutions to polynomial equations, including solutions in dimensions beyond three. (Solutions in two and three dimensions are first covered in plane and solid analytic geometry, respectively.)

Algebraic geometry emerged from analytic geometry after 1850 when topology, complex analysis, and algebra were used to study algebraic curves. An algebraic curve C is the graph of an equation $f(x, y) = 0$, with points at infinity added, where $f(x, y)$ is a polynomial, in two complex variables, that cannot be factored. Curves are classified by a nonnegative integer—known as their genus, g— that can be calculated from their polynomial.

The equation $f(x, y) = 0$ determines y as a function of x at all but a finite number of points of C. Since x takes values in the complex numbers, which are two-dimensional over the real numbers, the curve C is two-dimensional over the real numbers near most of its points. C looks like a hollow sphere with g hollow handles attached and finitely many points pinched together—a sphere has genus 0, a torus has genus 1, and so forth. The Riemann-Roch theorem uses integrals along paths on C to characterize g analytically.

A birational transformation matches up the points on two curves via maps given in both directions by rational functions of the coordinates. Birational transformations preserve intrinsic properties of curves, such as their genus, but provide leeway for geometers to simplify and classify curves by eliminating singularities (problematic points).

An algebraic curve generalizes to a variety, which is the solution set of r polynomial equations in n complex variables. In general, the difference n−r is the dimension of the variety—i.e., the number of independent complex parameters near most points. For example, curves have (complex) dimension one and surfaces have (complex) dimension two. The French mathematician Alexandre Grothendieck revolutionized algebraic geometry in the 1950s by generalizing varieties to schemes and extending the Riemann-Roch theorem.

DISCRETE GEOMETRY

Discrete geometry and combinatorial geometry are branches of geometry that study combinatorial properties and constructive methods of discrete geometric objects. Most questions in discrete geometry involve finite or discrete sets of basic geometric objects, such as points, lines, planes, circles, spheres, polygons, and so forth. The subject focuses on the combinatorial properties of these objects, such as how they intersect one another, or how they may be arranged to cover a larger object.

Discrete geometry has a large overlap with convex geometry and computational geometry, and is closely related to subjects such as finite geometry, combinatorial optimization, digital geometry, discrete differential geometry, geometric graph theory, toric geometry, and combinatorial topology.

Discrete Geometry

Polyhedra and Polytopes

A polytope is a geometric object with flat sides, which exists in any general number of dimensions. A polygon is a polytope in two dimensions, a polyhedron in three dimensions, and so on in higher dimensions (such as a 4-polytope in four dimensions). Some theories further generalize the idea to include such objects as unbounded polytopes (apeirotopes and tessellations), and abstract polytopes.

The following are some of the aspects of polytopes studied in discrete geometry:

- Polyhedral combinatorics,
- Lattice polytopes,
- Ehrhart polynomials,
- Pick's theorem,
- Hirsch conjecture.

Packings, Coverings and Tilings

Packings, coverings, and tilings are all ways of arranging uniform objects (typically circles, spheres, or tiles) in a regular way on a surface or manifold.

A sphere packing is an arrangement of non-overlapping spheres within a containing space. The spheres considered are usually all of identical size, and the space is usually three-dimensional Euclidean space. However, sphere packing problems can be generalised to consider unequal spheres, n-dimensional Euclidean space (where the problem becomes circle packing in two dimensions, or hypersphere packing in higher dimensions) or to non-Euclidean spaces such as hyperbolic space.

A tessellation of a flat surface is the tiling of a plane using one or more geometric shapes, called tiles, with no overlaps and no gaps. In mathematics, tessellations can be generalized to higher dimensions.

Specific topics in this area include:

- Circle packings,
- Sphere packings,
- Kepler conjecture,
- Quasicrystals,

- Aperiodic tilings,

- Periodic graph,

- Finite subdivision rules.

Structural Rigidity and Flexibility

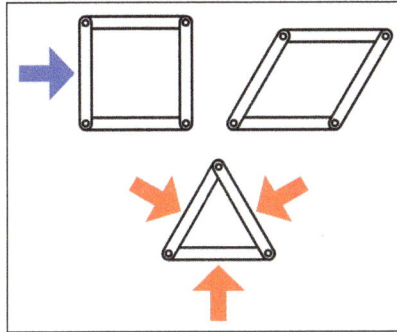

Graphs are drawn as rods connected by rotating hinges. The cycle graph C_4 drawn as a square can be tilted over by the blue force into a parallelogram, so it is a flexible graph. K_3, drawn as a triangle, cannot be altered by any force that is applied to it, so it is a rigid graph.

Structural rigidity is a combinatorial theory for predicting the flexibility of ensembles formed by rigid bodies connected by flexible linkages or hinges.

Topics in this area include:

- Cauchy's theorem.

- Flexible polyhedra.

Incidence Structures

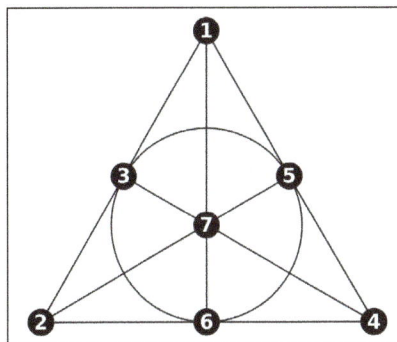

Seven points are elements of seven lines in the Fano plane, an example of an incidence structure.

Incidence structures generalize planes (such as affine, projective, and Möbius planes) as can be seen from their axiomatic definitions. Incidence structures also generalize the higher-dimensional analogs and the finite structures are sometimes called finite geometries.

Formally, an incidence structure is a triple:

$$C = (P, L, I).$$

where P is a set of "points", L is a set of "lines" and $I \subseteq P \times L$ is the incidence relation. The elements of I are called flags. If:

$$(p, l) \in I,$$

we say that point p "lies on" line l.

Topics in this area include:

- Configurations,
- Line arrangements,
- Hyperplane arrangements,
- Buildings.

Oriented Matroids

An oriented matroid is a mathematical structure that abstracts the properties of directed graphs and of arrangements of vectors in a vector space over an ordered field (particularly for partially ordered vector spaces). In comparison, an ordinary (i.e., non-oriented) matroid abstracts the dependence properties that are common both to graphs, which are not necessarily *directed*, and to arrangements of vectors over fields, which are not necessarily *ordered*.

Geometric Graph Theory

A geometric graph is a graph in which the vertices or edges are associated with geometric objects. Examples include Euclidean graphs, the 1-skeleton of a polyhedron or polytope, intersection graphs, and visibility graphs.

Topics in this area include:

- Graph drawing,
- Polyhedral graphs,
- Voronoi diagrams and Delaunay triangulations.

Simplicial Complexes

A simplicial complex is a topological space of a certain kind, constructed by "gluing together" points, line segments, triangles, and their n-dimensional counterparts. Simplicial complexes should not be confused with the more abstract notion of a simplicial set appearing in modern simplicial homotopy theory. The purely combinatorial counterpart to a simplicial complex is an abstract simplicial complex.

Topological Combinatorics

The discipline of combinatorial topology used combinatorial concepts in topology and in the early 20th century this turned into the field of algebraic topology.

In 1978, the situation was reversed – methods from algebraic topology were used to solve a problem in combinatorics – when László Lovász proved the Kneser conjecture, thus beginning the new study of topological combinatorics. Lovász's proof used the Borsuk-Ulam theorem and this theorem retains a prominent role in this new field. This theorem has many equivalent versions and analogs and has been used in the study of fair division problems.

Topics in this area include:

- Sperner's lemma.

- Regular maps.

Lattices and Discrete Groups

A discrete group is a group G equipped with the discrete topology. With this topology, G becomes a topological group. A discrete subgroup of a topological group G is a subgroup H whose relative topology is the discrete one. For example, the integers, Z, form a discrete subgroup of the reals, R (with the standard metric topology), but the rational numbers, Q, do not.

A lattice in a locally compact topological group is a discrete subgroup with the property that the quotient space has finite invariant measure. In the special case of subgroups of R^n, this amounts to the usual geometric notion of a lattice, and both the algebraic structure of lattices and the geometry of the totality of all lattices are relatively well understood. Deep results of Borel, Harish-Chandra, Mostow, Tamagawa, M. S. Raghunathan, Margulis, Zimmer obtained from the 1950s through the 1970s provided examples and generalized much of the theory to the setting of nilpotent Lie groups and semisimple algebraic groups over a local field. In the 1990s, Bass and Lubotzky initiated the study of *tree lattices*, which remains an active research area.

Topics in this area include:

- Reflection groups.

- Triangle groups.

Discrete Differential Geometry

Discrete differential geometry is the study of discrete counterparts of notions in differential geometry. Instead of smooth curves and surfaces, there are polygons, meshes, and simplicial complexes. It is used in the study of computer graphics and topological combinatorics.

Topics in this area include:

- Discrete Laplace operator.

- Discrete exterior calculus.

- Discrete Morse theory.

- Topological combinatorics.

- Spectral shape analysis.

- Abstract differential geometry.

- Analysis on fractals.

DIGITAL GEOMETRY

Digital geometry deals with discrete sets (usually discrete point sets) considered to be digitized models or images of objects of the 2D or 3D Euclidean space.

Simply put, digitizing is replacing an object by a discrete set of its points. The images we see on the TV screen, the raster display of a computer, or in newspapers are in fact digital images.

Its main application areas are computer graphics and image analysis.

Main aspects of study are:

- Constructing digitized representations of objects, with the emphasis on precision and efficiency (either by means of synthesis, see, for example, Bresenham's line algorithm or digital disks, or by means of digitization and subsequent processing of digital images).

- Study of properties of digital sets; see, for example, Pick's theorem, digital convexity, digital straightness, or digital planarity.

- Transforming digitized representations of objects, for example (A) into simplified shapes such as (i) skeletons, by repeated removal of simple points such that the digital topology of an image does not change, or (ii) medial axis, by calculating local maxima in a distance transform of the given digitized object representation, or (B) into modified shapes using mathematical morphology.

- Reconstructing "real" objects or their properties (area, length, curvature, volume, surface area, and so forth) from digital images.

- Study of digital curves, digital surfaces, and digital manifolds.

- Designing tracking algorithms for digital objects.

- Functions on digital space.

Digital geometry heavily overlaps with discrete geometry and may be considered as a part thereof.

Digital Space

A 2D digital space usually means a 2D grid space that only contains integer points in 2D Euclidean space. A 2D image is a function on a 2D digital space.

In Rosenfeld and Kak's book, digital connectivity are defined as the relationship among elements in digital space. For example, 4-connectivity and 8-connectivity in 2D. A digital space and its (digital-)connectivity determine a digital topology.

In digital space, the digitally continuous function and the gradually varied function were proposed, independently.

A digitally continuous function means a function in which the value (an integer) at a digital point is the same or off by at most 1 from its neighbors. In other words, if x and y are two adjacent points in a digital space, $|f(X) - f(y)| \le 1$.

A gradually varied function is a function from a digital space Σ to $\{A_1, ..., A_m\}$ where $A_1 \ll \cdots < A_m$ and A_i are real numbers. This function possesses the following property: If x and y are two adjacent points in Σ, assume $f_{(x)} = A_i$, then $f_{(y)} = A_i$, $f_{(x)} = A_{i+1}$, or A_{i-1}. So we can see that the gradually varied function is defined to be more general than the digitally continuous function.

An extension theorem related to above functions was mentioned by A. Rosenfeld and completed by L. Chen. This theorem states: Let $D \subset \Sigma$ and $f = D \to \{A_1, ..., A_m\}$. The necessary and sufficient condition for the existence of the gradually varied extension F of f is : for each pair of points x and y in D, assume $f_{(x)} = A_i$ and $f_{(y)} = A_j$, we have $|i - j| \le d(x, y)$, where $d(x, y)$ is the (digital) distance between x and y.

FINITE GEOMETRY

A finite geometry is any geometric system that has only a finite number of points. The familiar Euclidean geometry is not finite, because a Euclidean line contains infinitely many points. A geometry based on the graphics displayed on a computer screen, where the pixels are considered to be the points, would be a finite geometry. While there are many systems that could be called finite geometries, attention is mostly paid to the finite projective and affine spaces because of their regularity and simplicity. Other significant types of finite geometry are finite Möbius or inversive planes and Laguerre planes, which are examples of a general type called Benz planes, and their higher-dimensional analogs such as higher finite inversive geometries.

Finite geometries may be constructed via linear algebra, starting from vector spaces over a finite field; the affine and projective planes so constructed are called Galois geometries. Finite geometries can also be defined purely axiomatically. Most common finite geometries are Galois geometries, since any finite projective space of dimension three or greater is isomorphic to a projective space over a finite field (that is, the projectivization of a vector space over a finite field). However, dimension two has affine and projective planes that are not isomorphic to Galois geometries, namely the non-Desarguesian planes. Similar results hold for other kinds of finite geometries.

Finite Planes

The following remarks apply only to finite *planes*. There are two main kinds of finite plane geometry: affine and projective. In an affine plane, the normal sense of parallel lines applies. In a projective plane, by contrast, any two lines intersect at a unique point, so parallel lines do not exist. Both finite affine plane geometry and finite projective plane geometry may be described by fairly simple axioms.

Finite Affine Planes

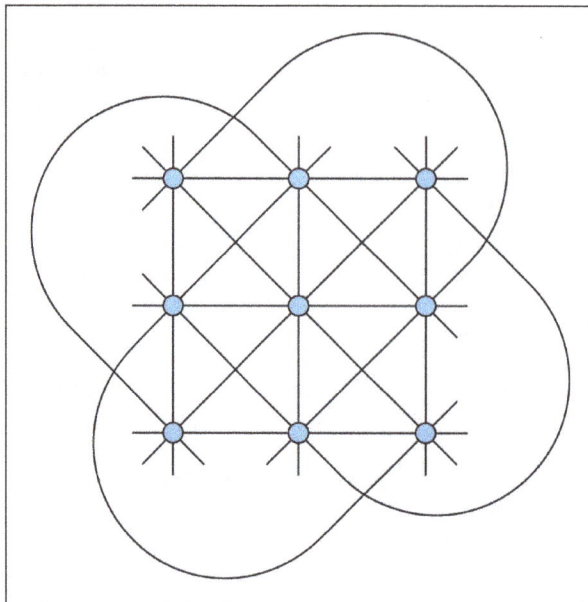

Finite affine plane of order 3, containing 9 points and 12 lines.

An affine plane geometry is a nonempty set X (whose elements are called "points"), along with a nonempty collection L of subsets of X (whose elements are called "lines"), such that:

1. For every two distinct points, there is exactly one line that contains both points.

2. Playfair's axiom: Given a line ℓ and a point p not on ℓ, there exists exactly one line ℓ' containing p such that $\ell' \cap \ell' = \varnothing$.

3. There exists a set of four points, no three of which belong to the same line.

The last axiom ensures that the geometry is not *trivial* (either empty or too simple to be of interest, such as a single line with an arbitrary number of points on it), while the first two specify the nature of the geometry.

The simplest affine plane contains only four points; it is called the affine plane of order 2. (The order of an affine plane is the number of points on any line.) Since no three are collinear, any pair of points determines a unique line, and so this plane contains six lines. It corresponds to a tetrahedron where non-intersecting edges are considered "parallel", or a square where not only opposite sides, but also diagonals are considered "parallel". More generally, a finite affine plane of order n has n^2 points and $n^2 + n$ lines; each line contains n points, and each point is on $n + 1$ lines. The affine plane of order 3 is known as the Hesse configuration.

Finite Projective Planes

A projective plane geometry is a nonempty set X (whose elements are called "points"), along with a nonempty collection L of subsets of X (whose elements are called "lines"), such that,

1. For every two distinct points, there is exactly one line that contains both points.

2. The intersection of any two distinct lines contains exactly one point.

3. There exists a set of four points, no three of which belong to the same line.

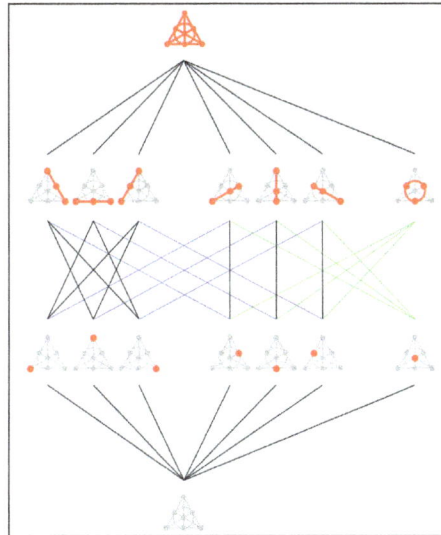

Duality in the Fano plane: Each point corresponds to a line and vice versa.

An examination of the first two axioms shows that they are nearly identical, except that the roles of points and lines have been interchanged. This suggests the principle of duality for projective plane geometries, meaning that any true statement valid in all these geometries remains true if we exchange points for lines and lines for points. The smallest geometry satisfying all three axioms contains seven points. In this simplest of the projective planes, there are also seven lines; each point is on three lines, and each line contains three points.

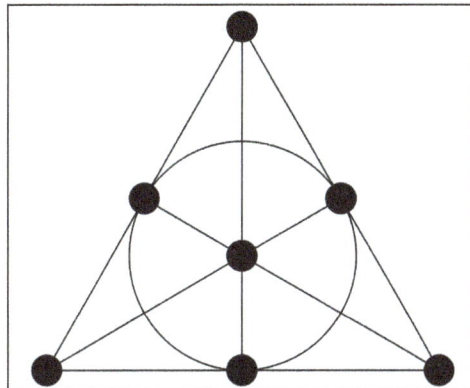

The Fano plane.

This particular projective plane is sometimes called the Fano plane. If any of the lines is removed from the plane, along with the points on that line, the resulting geometry is the affine plane of order 2. The Fano plane is called the *projective plane of order* 2 because it is unique (up to isomorphism). In general, the projective plane of order n has $n^2 + n + 1$ points and the same number of lines; each line contains $n + 1$ points, and each point is on $n + 1$ lines.

A permutation of the Fano plane's seven points that carries collinear points (points on the same line) to collinear points is called a collineation of the plane. The full collineation group is of order

168 and is isomorphic to the group PSL(2,7) ≈ PSL(3,2), which in this special case is also isomorphic to the general linear group GL(3,2) ≈ PGL(3,2).

Order of Planes

A finite plane of order n is one such that each line has n points (for an affine plane), or such that each line has $n + 1$ points (for a projective plane). One major open question in finite geometry is:

> "Is the order of a finite plane always a prime power?"

This is conjectured to be true.

Affine and projective planes of order n exist whenever n is a prime power (a prime number raised to a positive integer exponent), by using affine and projective planes over the finite field with $n = p^k$ elements. Planes not derived from finite fields also exist, but all known examples have order a prime power.

The best general result to date is the Bruck–Ryser theorem of 1949, which states:

> "If n is a positive integer of the form $4k + 1$ or $4k + 2$ and n is not equal to the sum of two integer squares, then n does not occur as the order of a finite plane".

The smallest integer that is not a prime power and not covered by the Bruck–Ryser theorem is 10; 10 is of the form $4k + 2$, but it is equal to the sum of squares $1^2 + 3^2$. The non-existence of a finite plane of order 10 was proven in a computer-assisted proof that finished in 1989.

The next smallest number to consider is 12, for which neither a positive nor a negative result has been proved.

Finite Spaces of 3 or more Dimensions

For some important differences between finite *plane* geometry and the geometry of higher-dimensional finite spaces. For a discussion of higher-dimensional finite spaces in general, see, for instance, the works of J.W.P. Hirschfeld. The study of these higher-dimensional spaces ($n \geq 3$) has many important applications in advanced mathematical theories.

Axiomatic

A projective space S can be defined axiomatically as a set P (the set of points), together with a set L of subsets of P (the set of lines), satisfying these axioms:

- Each two distinct points p and q are in exactly one line.

- Veblen's axiom: If a, b, c, d are distinct points and the lines through ab and cd meet, then so do the lines through ac and bd.

- Any line has at least 3 points on it.

The last axiom eliminates reducible cases that can be written as a disjoint union of projective spaces together with 2-point lines joining any two points in distinct projective spaces. More abstractly, it can be defined as an incidence structure (P, L, I) consisting of a set P of points, a set L of lines, and an incidence relation I stating which points lie on which lines.

Obtaining a *finite* projective space requires one more axiom:

- The set of points P is a finite set.

In any finite projective space, each line contains the same number of points and the *order* of the space is defined as one less than this common number.

A subspace of the projective space is a subset X, such that any line containing two points of X is a subset of X (that is, completely contained in X). The full space and the empty space are always subspaces.

The geometric dimension of the space is said to be n if that is the largest number for which there is a strictly ascending chain of subspaces of this form:

$$\varnothing = X_{-1} \subset X_0 \subset \cdots \subset X_n = P.$$

Algebraic Construction

A standard algebraic construction of systems satisfies these axioms. For a division ring D construct an $(n + 1)$-dimensional vector space over D (vector space dimension is the number of elements in a basis). Let P be the 1-dimensional (single generator) subspaces and L the 2-dimensional (two independent generators) subspaces (closed under vector addition) of this vector space. Incidence is containment. If D is finite then it must be a finite field GF(q), since by Wedderburn's little theorem all finite division rings are fields. In this case, this construction produces a finite projective space. Furthermore, if the geometric dimension of a projective space is at least three then there is a division ring from which the space can be constructed in this manner. Consequently, all finite projective spaces of geometric dimension at least three are defined over finite fields. A finite projective space defined over such a finite field has $q + 1$ points on a line, so the two concepts of order coincide. Such a finite projective space is denoted by PG(n, q), where PG stands for projective geometry, n is the geometric dimension of the geometry and q is the size (order) of the finite field used to construct the geometry.

In general, the number of k-dimensional subspaces of PG(n, q) is given by the product:

$$\binom{n+1}{k+1}_q = \prod_{i=0}^{k} \frac{q^{n+1-i} - 1}{q^{i+1} - 1},$$

which is a Gaussian binomial coefficient, a q analogue of a binomial coefficient.

Classification of Finite Projective Spaces by Geometric Dimension

- Dimension 0 (no lines): The space is a single point and is so degenerate that it is usually ignored.
- Dimension 1 (exactly one line): All points lie on the unique line, called a projective line.

- Dimension 2: There are at least 2 lines, and any two lines meet. A projective space for n = 2 is a projective plane. These are much harder to classify, as not all of them are isomorphic with a PG(d, q). The Desarguesian planes (those that are isomorphic with a PG(2, q)) satisfy Desargues's theorem and are projective planes over finite fields, but there are many non-Desarguesian planes.

- Dimension at least 3: Two non-intersecting lines exist. The Veblen–Young theorem states in the finite case that every projective space of geometric dimension $n \geq 3$ is isomorphic with a PG(n, q), the n-dimensional projective space over some finite field GF(q).

The Smallest Projective Three-space

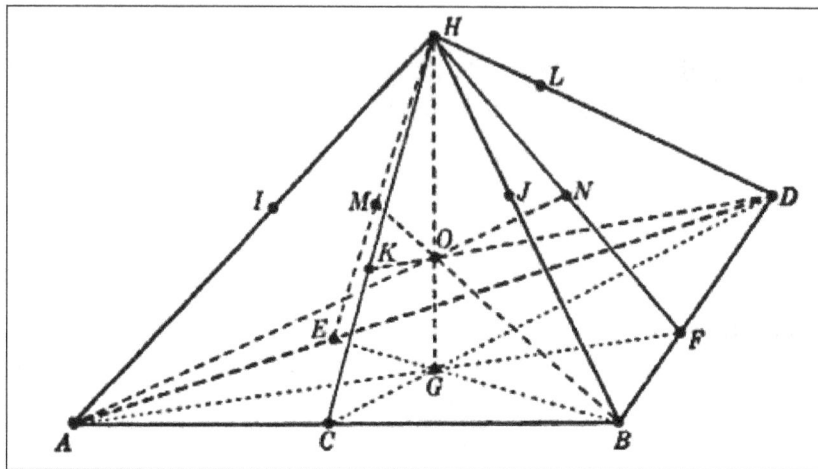

PG(3,2) but not all the lines are drawn.

The smallest 3-dimensional projective space is over the field GF(2) and is denoted by PG(3,2). It has 15 points, 35 lines, and 15 planes. Each plane contains 7 points and 7 lines. Each line contains 3 points. As geometries, these planes are isomorphic to the Fano plane.

Every point is contained in 7 lines. Every pair of distinct points are contained in exactly one line and every pair of distinct planes intersects in exactly one line.

In 1892, Gino Fano was the first to consider such a finite geometry.

Kirkman's Schoolgirl Problem

PG(3,2) arises as the background for a solution of Kirkman's schoolgirl problem, which states: "Fifteen schoolgirls walk each day in five groups of three. Arrange the girls' walk for a week so that in that time, each pair of girls walks together in a group just once." There are 35 different combinations for the girls to walk together. There are also 7 days of the week, and 3 girls in each group. Two of the seven non-isomorphic solutions to this problem can be stated in terms of structures in the Fano 3-space, PG(3,2), known as *packings*. A *spread* of a projective space is a partition of its points into disjoint lines, and a packing is a partition of the lines into disjoint spreads. In PG(3,2), a spread would be a partition of the 15 points into 5 disjoint lines (with 3 points on each line), thus corresponding to the arrangement of schoolgirls on a particular day. A packing of PG(3,2) consists of seven disjoint spreads and so corresponds to a full week of arrangements.

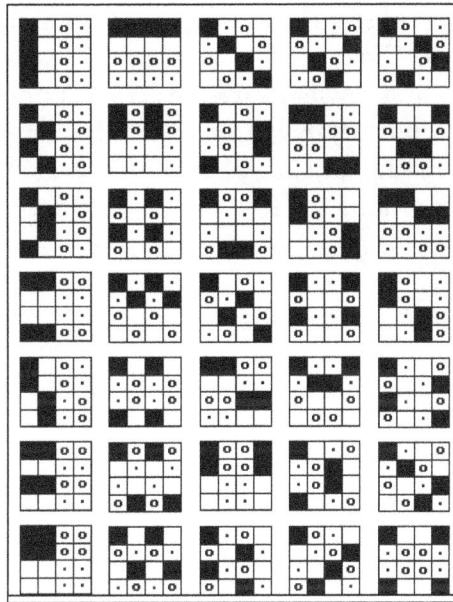

Square model of Fano 3-space.

ANALYTIC GEOMETRY

Analytic geometry is the mathematical subject in which algebraic symbolism and methods are used to represent and solve problems in geometry. The importance of analytic geometry is that it establishes a correspondence between geometric curves and algebraic equations. This correspondence makes it possible to reformulate problems in geometry as equivalent problems in algebra, and vice versa; the methods of either subject can then be used to solve problems in the other. For example, computers create animations for display in games and films by manipulating algebraic equations.

Elementary Analytic Geometry

Apollonius of Perga, known by his contemporaries as the "Great Geometer," foreshadowed the development of analytic geometry by more than 1,800 years with his book Conics. He defined a conic as the intersection of a cone and a plane. Using Euclid's results on similar triangles and on secants of circles, he found a relation satisfied by the distances from any point P of a conic to two perpendicular lines, the major axis of the conic and the tangent at an endpoint of the axis. These distances correspond to coordinates of P, and the relation between these coordinates corresponds to a quadratic equation of the conic. Apollonius used this relation to deduce fundamental properties of conics.

Further development of coordinate systems in mathematics emerged only after algebra had matured under Islamic and Indian mathematicians. At the end of the 16th century, the French mathematician François Viète introduced the first systematic algebraic notation, using letters to represent known and unknown numerical quantities, and he developed powerful general methods for working with algebraic expressions and solving algebraic equations. With the power of algebraic notation, mathematicians were no longer completely dependent upon geometric figures and geometric intuition to solve problems. The more daring began to leave behind the standard geometric way of thinking in which linear (first power) variables corresponded to lengths, squares (second power) to areas, and cubics (third power) to volumes, with higher powers lacking "physical" interpretation. Two Frenchmen, the mathematician-philosopher René Descartes and the lawyer-mathematician Pierre de Fermat, were among the first to take this daring step.

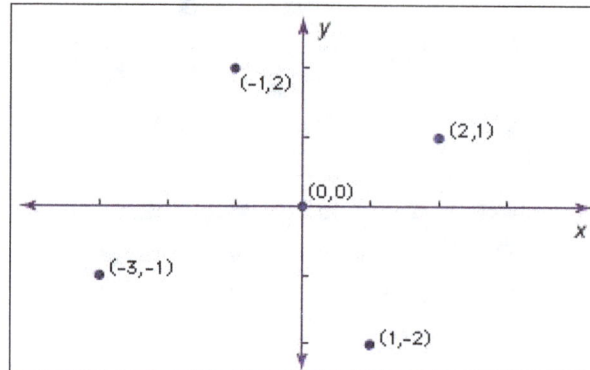

Cartesian coordinatesSeveral points are labeled in a two-dimensional graph, known as the Cartesian plane. Note that each point has two coordinates, the first number (x value) indicates its distance from the y-axis—positive values to the right and negative values to the left—and the second number (y value) gives its distance from the x-axis—positive values upward and negative values downward.

Descartes and Fermat independently founded analytic geometry in the 1630s by adapting Viète's algebra to the study of geometric loci. They moved decisively beyond Viète by using letters to represent distances that are variable instead of fixed. Descartes used equations to study curves defined geometrically, and he stressed the need to consider general algebraic curves—graphs of polynomial equations in x and y of all degrees. He demonstrated his method on a classical problem: finding all points P such that the product of the distances from P to certain lines equals the product of the distances to other lines.

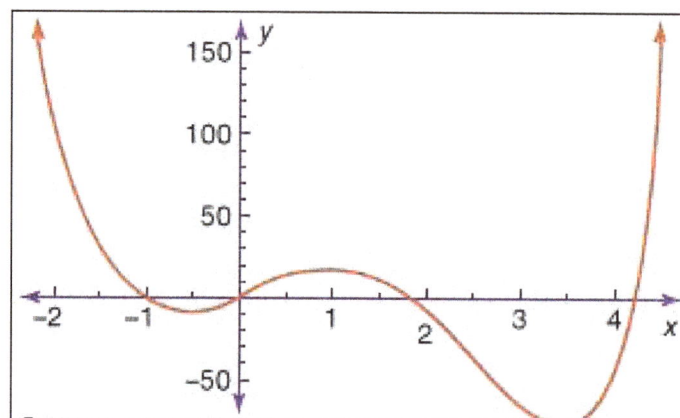

Polynomial graphThe figure shows part of the graph of the polynomial
equation $y = 3x^4 - 16x^3 + 6x^2 + 24x + 1$. Note that the same scale need not be used for the x- and y-axis.

Fermat emphasized that any relation between x and y coordinates determines a curve. Using this idea, he recast Apollonius's arguments in algebraic terms and restored lost work. Fermat indicated that any quadratic equation in x and y can be put into the standard form of one of the conic sections.

Fermat did not publish his work, and Descartes deliberately made his hard to read in order to discourage "dabblers." Their ideas gained general acceptance only through the efforts of other mathematicians in the latter half of the 17th century. He added vital explanatory material, as did the French lawyer Florimond de Beaune, and the Dutch mathematician Johan de Witt. In England, the mathematician John Wallis popularized analytic geometry, using equations to define conics and derive their properties. He used negative coordinates freely, although it was Isaac Newton who unequivocally used two (oblique) axes to divide the plane into four quadrants, as shown in the figure.

Analytic geometry had its greatest impact on mathematics via calculus. Without access to the power of analytic geometry, classical Greek mathematicians such as Archimedes solved special cases of the basic problems of calculus: finding tangents and extreme points (differential calculus) and arc lengths, areas, and volumes (integral calculus). Renaissance mathematicians were led back to these problems by the needs of astronomy, optics, navigation, warfare, and commerce. They naturally sought to use the power of algebra to define and analyze a growing range of curves.

Fermat developed an algebraic algorithm for finding the tangent to an algebraic curve at a point by finding a line that has a double intersection with the curve at the point—in essence, inventing differential calculus. Descartes introduced a similar but more complicated algorithm using a circle. Fermat computed areas under the curves $y = ax^k$ for all rational numbers $k \neq -1$ by summing areas of inscribed and circumscribed rectangles. For the rest of the 17th century, the groundwork for calculus was continued by many mathematicians, including the Frenchman Gilles Personne de Roberval, the Italian Bonaventura Cavalieri, and the Britons James Gregory, John Wallis, and Isaac Barrow.

Newton and the German Gottfried Leibniz revolutionized mathematics at the end of the 17th century by independently demonstrating the power of calculus. Both men used coordinates to develop notations that expressed the ideas of calculus in full generality and led naturally to differentiation rules and the fundamental theorem of calculus (connecting differential and integral calculus).

Newton demonstrated the importance of analytic methods in geometry, apart from their role in calculus, when he asserted that any cubic—or, algebraic curve of degree three—has one of four standard equations,

$$xy^2 + ey = ax^3 + bx^2 + cx + d,$$

$$xy = ax^3 + bx^2 + cx + d,$$

$$y^2 = ax^3 + bx^2 + cx + d,$$

$$y = ax^3 + bx^2 + cx + d,$$

for suitable coordinate axes. The Scottish mathematician James Stirling proved this assertion in 1717, possibly with Newton's aid. Newton divided cubics into 72 species, a total later corrected to 78.

Newton also showed how to express an algebraic curve near the origin in terms of the fractional power series $y = a_1x^{1/k} + a_2x^{2/k} + \ldots$ for a positive integer k. Mathematicians have since used this technique to study algebraic curves of all degrees.

Analytic Geometry of Three and more Dimensions

Although both Descartes and Fermat suggested using three coordinates to study curves and surfaces in space, three-dimensional analytic geometry developed slowly until about 1730, when the Swiss mathematicians Leonhard Euler and Jakob Hermann and the French mathematician Alexis Clairaut produced general equations for cylinders, cones, and surfaces of revolution. For example, Euler and Hermann showed that the equation $f(z) = x^2 + y^2$ gives the surface that is produced by revolving the curve $f(z) = x^2$ about the z-axis.

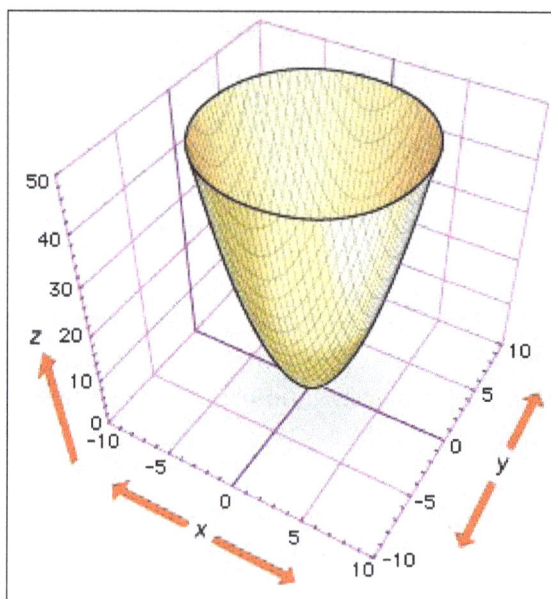

Elliptic paraboloid The figure shows part of the elliptic paraboloid $z = x^2 + y^2$, which can be generated by rotating the parabola $z = x^2$ (or $z = y^2$) about the z-axis. Note that cross sections of the surface parallel to the xy plane, as shown by the cutoff at the top of the figure, are ellipses.

Newton made the remarkable claim that all plane cubics arise from those in his third standard form by projection between planes. This was proved independently in 1731 by Clairaut and the French mathematician François Nicole. Clairaut obtained all the cubics in Newton's four standard forms as sections of the cubical cone:

$$zy^2 = ax^3 + bx^2z + cxz^2 + dz^3$$

consisting of the lines in space that join the origin (0, 0, 0) to the points on the third standard cubic in the plane $z = 1$.

In 1748 Euler used equations for rotations and translations in space to transform the general quadric surface:

$$ax^2 + by^2 + cz^2 + dxy + exz + fyz + gx + hy + iz + j = 0$$

so that its principal axes coincide with the coordinate axes. Euler and the French mathematicians

Joseph-Louis Lagrange and Gaspard Monge made analytic geometry independent of synthetic (nonanalytic) geometry.

Vector Analysis

In Euclidean space of any dimension, vectors—directed line segments—can be specified by coordinates. An n-tuple $(a_1, ..., a_n)$ represents the vector in n-dimensional space that projects onto the real numbers $a_1, ..., a_n$ on the coordinate axes.

In 1843 the Irish mathematician-astronomer William Rowan Hamilton represented four-dimensional vectors algebraically and invented the quaternions, the first noncommutative algebra to be extensively studied. Multiplying quaternions with one coordinate zero led Hamilton to discover fundamental operations on vectors. Nevertheless, mathematical physicists found the notation used in vector analysis more flexible—in particular, it is readily extendable to infinite-dimensional spaces. The quaternions remained of interest algebraically and were incorporated in the 1960s into certain new particle physics models.

Projections

As readily available computing power grew exponentially in the last decades of the 20th century, computer animation and computer-aided design became ubiquitous. These applications are based on three-dimensional analytic geometry. Coordinates are used to determine the edges or parametric curves that form boundaries of the surfaces of virtual objects. Vector analysis is used to model lighting and determine realistic shadings of surfaces.

As early as 1850, Julius Plücker had united analytic and projective geometry by introducing homogeneous coordinates that represent points in the Euclidean plane and at infinity in a uniform way as triples. Projective transformations, which are invertible linear changes of homogeneous coordinates, are given by matrix multiplication. This lets computer graphics programs efficiently change the shape or the view of pictured objects and project them from three-dimensional virtual space to the two-dimensional viewing screen.

COMPLEX GEOMETRY

In mathematics, complex geometry is the study of complex manifolds and functions of several complex variables. Application of transcendental methods to algebraic geometry falls in this category, together with more geometric aspects of complex analysis.

An *analytic subset* of a complex-analytic manifold M is locally the zero-locus of some family of holomorphic functions on M. It is called an analytic subvariety if it is irreducible in the Zariski topology.

Line Bundles and Divisors

Throughout this section, X denotes a complex manifold. Accordance with the definitions of the paragraph "line bundles and divisors" in "projective varieties", let the regular functions on X be denoted $\mathcal{O}$ and its invertible subsheaf $\mathcal{O}^*$. And let $\mathcal{M}_X$ be the sheaf on X associated with $U \mapsto$ the total ring of fractions of, $\Gamma(U, \mathcal{O}_X)$ where U_i are the open affine charts. Then a global section of $\mathcal{M}_X^* / \mathcal{O}_X^*$ (* means multiplicative group) is called a Cartier divisor on X.

Let $Pic(X)$ be the set of all isomorphism classes of line bundles on X. It is called the Picard group of X and is naturally isomorphic to $H^1(X, \mathcal{O}^*)$. Taking the short exact sequence of:

$$0 \to \mathbb{z} \to \mathcal{O} \to \mathcal{O}^* \to 1$$

where the second map is $f \mapsto \exp(2\pi i f)$ yields a homomorphism of groups:

$$Pic(X) \to H^2(X, \mathbb{z}).$$

The image of a line bundle $\mathcal{L}$ under this map is denoted by $c_1(\mathcal{L})$ and is called the first Chern class of $\mathcal{L}$.

A divisor D on X is a formal sum of hypersurfaces (subvariety of codimension one):

$$D = \sum a_i V_i, \quad a_i \in \mathbb{z}$$

that is locally a finite sum. The set of all divisors on X is denoted by $Div(X)$. It can be canonically identified with $H^0(X, \mathcal{M}^* / \mathcal{O}^*)$. Taking the long exact sequence of the quotient $\mathcal{M}^* / \mathcal{O}^*$, one obtains a homomorphism:

$$Div(X) \to Pic(X).$$

A line bundle is said to be positive if its first Chern class is represented by a closed positive real $(1,1)$-form. Equivalently, a line bundle is positive if it admits a hermitian structure such that the induced connection has Griffiths-positive curvature. A complex manifold admitting a positive line bundle is kähler.

The Kodaira embedding theorem states that a line bundle on a compact kähler manifold is positive if and only if it is ample.

Complex Vector Bundles

Let X be a differentiable manifold. The basic invariant of a complex vector bundle $\pi: E \to X$ is the Chern class of the bundle. By definition, it is a sequence $c_1, c_2, \ldots$ such that $c_i(E)$ is an element of $H^{2i}(X, \mathbb{z})$ and that satisfies the following axioms:

1. $c_i(f^*(E)) = f^*(c_i(E))$ for any differentiable map $f: Z \to X$.

2. $c(E \oplus F) = c(E) \cup c(F)$ where F is another bundle and $c = 1 + c_1 + c_2 + \ldots$.

3. $c_i(E) = 0$ for $i > \mathrm{rk}\, E$.

4. $-c_1(E_1)$ generates $H^2(\mathbb{c}\mathbf{P}^1, \mathbb{z})$ where E_1 is the canonical line bundle over $\mathbb{c}\mathbf{P}^1$.

If L is a line bundle, then the Chern character of L is given by:

$$\mathrm{ch}(L) = e^{c_1(L)}.$$

More generally, if E is a vector bundle of rank r, then we have the formal factorization:

$$\sum c_i(E)t^i = \prod_1^r (1+\eta_i t) \text{ and then we set.}$$

$$\text{ch}(E) = \sum e^{\eta_i}.$$

GEOMETRY APPLICATIONS: ART, SCIENCE AND EVERYDAY LIFE

Geometry in Art

Art is a difficult word to define because it means different things to many people. Art can be visual, auditory, movement or something else. One thing that everyone can say is true is geometry can be seen in all forms of art. Let's look at some examples of this.

The study of geometry starts with single points and branches out to lines and then to three dimensional objects. If you have ever seen a painting or picture of the night sky you saw stars, which are single points. The arrangement of the stars relative to each other allowed the ancient Greeks to connect some stars with lines to form basic shapes, which are constellations.

Connecting stars with lines forming constellations.

The ancient Egyptians decorated the Giza plateau with enormous pyramids made of stone blocks. The ancient Maya people in what is currently Mexico and Central America made more of a cubic style of pyramid. A modern artist designed a pyramid made partly of glass marking the location of the famous art museum the Louvre in Paris, France.

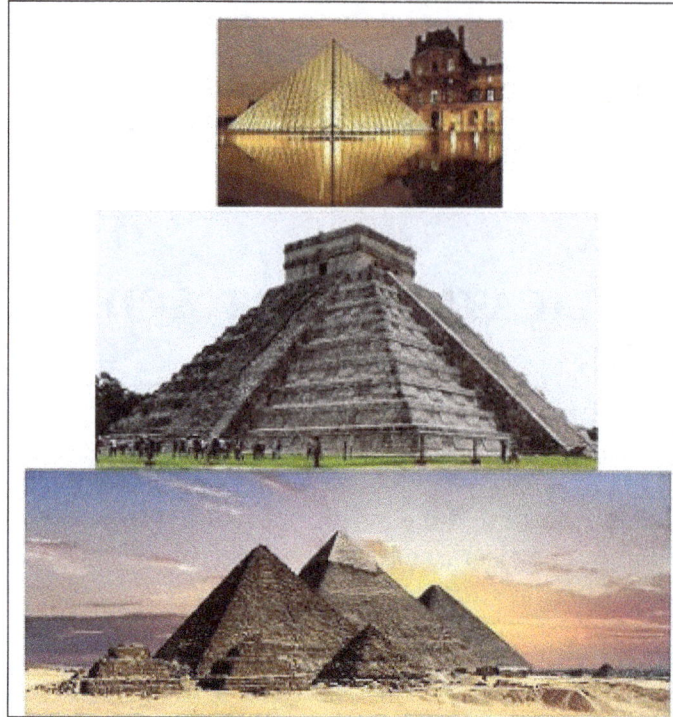

Pyramids: Louvre at top, Mayan temple in middle, and Egyptian pyramids on bottom.

The ancient Egyptians also made obelisks as commemoratory markers for people deserving recognition, special events, or simply to honor their gods. An obelisk is a rectangular prism that is tapered towards the top and is capped with a pyramid.

The Washington monument is an obelisk.

Let's say an artist drew or painted a scene of a road leading into the distance. Artists know about perspective, and when they draw a road or pathway seemingly heading into the distance they draw two lines that start relatively far apart and have them look like they converge in the distance. Let's take our own path from art to science.

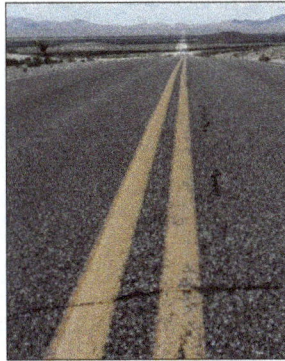

Parallel lines that look like they converge in the distance.

Geometry in Science

When you learned about spheres in geometry, were you thinking you were sitting on a giant sphere called Earth? The same equation used to calculate the volume, or surface area of a small plastic sphere is used to determine the volume and surface area of planets moons and stars.

When dealing with electricity and magnetism in physics, imaginary shapes are generated around charged objects called Gaussian surfaces. The use of geometric shapes makes the math much more manageable when determining electric or magnetic flux, which is the amount of electric or magnetic field that penetrates an area.

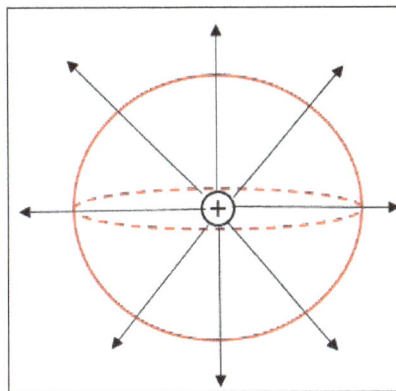

Gaussian sphere shown in red. Arrows represent electric field.

In chemistry, spherical shapes describe certain orbitals where electrons can be located in the electron cloud. Covalent molecules have specific shapes based on the arrangement of electrons in the elements that comprise them. Since atoms and molecules make up everything that physically exists let's branch out into everyday life and see what geometric applications we can find.

Geometry in Everyday Life

Geometry in everyday life deals with function or appearance. This is science and art combined! Let's put on those special glasses we talked about previously and take a look around.

The first thing you probably see is your computer screen, which is rectangular. When manufacturers report the size of computer screens and televisions they are reporting the measurement across their diagonals.

References

- What-is-geometry, what-is-pure-math, about-pure-math, pure-mathematics: uwaterloo.ca, Retrieved 19 April, 2019

- Differential-geometry, science: britannica.com, Retrieved 21 May, 2019

- bárány, imre (2010), "discrete and convex geometry", in horváth, jános (ed.), a panorama of hungarian mathematics in the twentieth century, i, new york: springer, pp. 431–441, isbn 9783540307211

- Algebraic-geometry, science: britannica.com, Retrieved 8 January, 2019

- Ball, simeon (2015), finite geometry and combinatorial applications, london mathematical society student texts, cambridge university press, isbn 978-1107518438

- Analytic-geometry, science: britannica.com Retrieved 13 May, 2019

- Geometry-applications-art-science-everyday-life, lesson, academy: study.com, Retrieved 6 January, 2019

2

Fundamental Concepts of Geometry

Some of the basic elements of geometry are point, lines, acute angle, obtuse angle, right angle, whole angle, etc. The topics elaborated in this chapter will help in gaining a better perspective about these elements of geometry.

POINTS

In geometry, we use points to specify exact locations. They are generally denoted by a number or letter. Because points specify a single, exact location, they are zero-dimensional. In other words, points have no length, width, or height. It may be helpful to think of a point as a miniscule "dot" on a piece of paper.

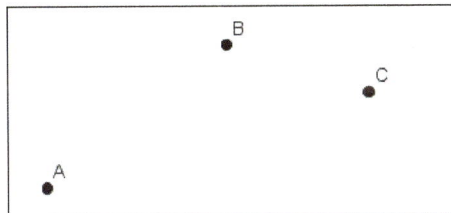

Three geometric points labeled A, B, and C.

Lines

Lines in geometry may be thought of as a "straight" line that can be drawn on paper with a pencil and ruler. However, instead of this line being bounded by the dimensions of the paper, a line extends infinitely in both directions. A line is one-dimensional, having length, but no width or height. Lines are uniquely determined by two points. Thus, we denote the name of a line passing through the points A and B as, where the two-headed arrow signifies that the line passes through those unique points and extends infinitely in both directions.

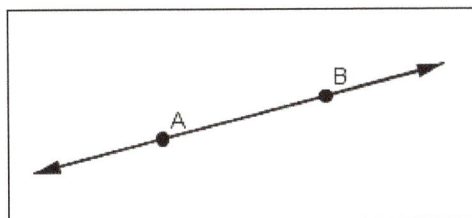

An infinitely extending line with two points, labeled A and B.

30

Geometry: A Comprehensive Course

Line Segments

Consider the task of drawing a "straight" line on a piece of paper (as we've done when thinking about lines). What you've actually done is create a line segment. Because our piece of paper has defined dimensions and we cannot draw a line infinitely in any direction, we have constructed a segment that begins somewhere and ends somewhere. We write the name of a line segment with endpoints A and B as. Note that the notation for lines and line segments differ because a line segment has a defined length, whereas a line does not.

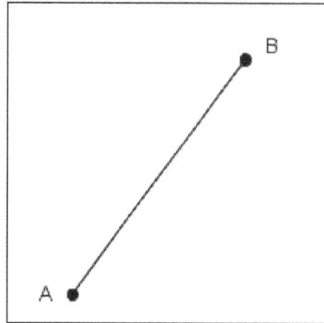

A line segment with two endpoints labeled A and B.

Rays

A ray is a "straight" line that begins at a certain point and extends infinitely in one direction. A ray has one endpoint, which marks the position from where it begins. A ray beginning at the point A that passes through point B is denoted as $\overrightarrow{AB}$. This notation shows that the ray begins at point A and extends infinitely in the direction of point B.

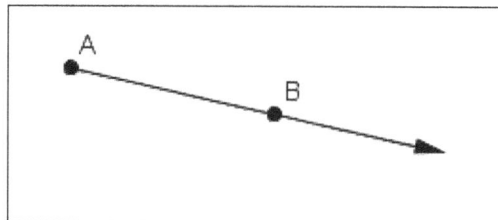

A ray that extends infinitely in one direction, beginning at point A with another point labeled B.

Endpoints

Endpoints mark the beginning or end of a line segment or ray. Line segments have two endpoints, giving them defined lengths, whereas rays only have one endpoint, so the length of a ray cannot be measured.

Midpoints

The midpoint of a line segment marks the point at which the segment is divided into two equal segments. In other words, the lengths of the segments from either endpoint to the midpoint are equal. For instance, if M is the midpoint of the segment $\overline{AB}$, then $\overline{AM} = \overline{AB}$. Note that neither lines nor rays can have midpoints because they extend infinitely in at least one direction. It would be impossible to find the middle of a line or ray that never ends.

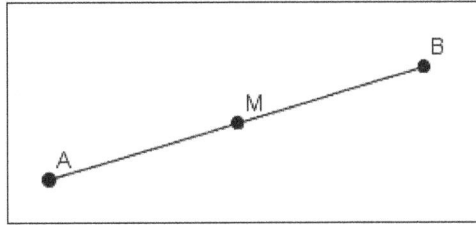

A line segment with endpoints labeled A and B, and a midpoint labeled M.

Intersection

When we have lines, line segments, or rays that meet, or cross at a certain point, we call it an intersection point. In other words, those figures intersect somewhere.

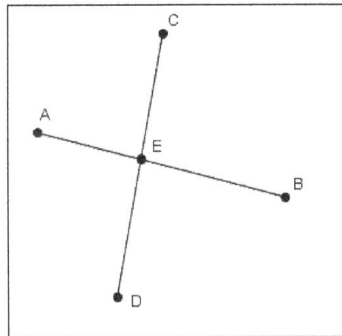

Two lines intersecting, with an intersection point labeled E.

Parallel

Two lines that will never intersect are called parallel lines. In the case of line segments and rays, we must consider the lines that they lie in. In other words, we must consider the case that the line segments or rays were actually lines that extend infinitely in both directions. If the lines they lie on never intersect, they are called parallel. For instance, the statement "$\overline{AB}$ is parallel to $\overline{CD}$," is expressed mathematically as $\overline{AB} \parallel \overline{CD}$.

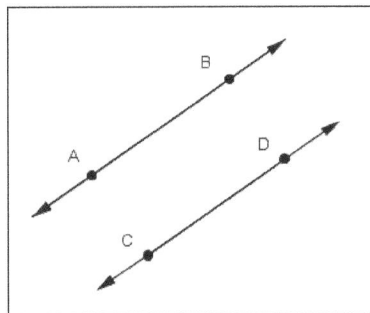

Two parallel lines with points labeled A, B, C, and D
If extended infinitely, the lines above will never meet.

Transversal

A transversal is a type of line that intersects at least two other lines. The lines that a transversal crosses may or may not be parallel.

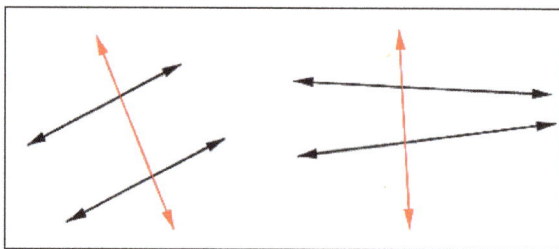

Two examples of transversals highlighted red, intersecting parallel and non-parallel linesIn both figures, the red line is a transversal.

Planes

A plane can be thought of as a two-dimensional flat surface, having length and width, but no height. A plane extends indefinitely on all sides and is composed of an infinite number of points and lines. One way to think about a plane is as a sheet of paper with infinite length and width.

Space

Space is the set of all possible points on an infinite number of planes. Thus, space covers all three dimensions - length, width, and height.

ANGLE

In geometry and trigonometry, an angle (or plane angle) is the figure formed by two rays sharing a common endpoint. The endpoint is called the vertex of the angle. The magnitude of the angle is the "amount of rotation" that separates the two rays, and can be measured by considering the length of circular arc swept out when one ray is rotated about the vertex to coincide with the other.

Measuring Angles

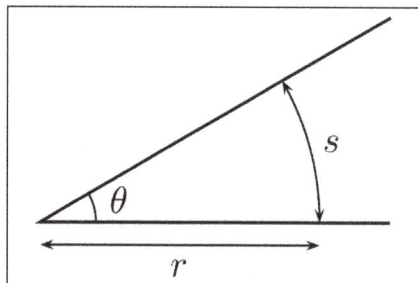

The angle θ is the quotient of s and r.

In order to measure an angle θ, a circular arc centered at the vertex of the angle is drawn, e.g., with a pair of compasses. The length of the arc s is then divided by the radius of the circle r, and possibly multiplied by a scaling constant k (which depends on the units of measurement that are chosen):

$$\theta = \frac{s}{r}(k)$$

The value of θ thus defined is independent of the size of the circle: if the length of the radius is changed then the arc length changes in the same proportion, so the ratio s/r is unaltered.

In many geometrical situations, angles that differ by an exact multiple of a full circle are effectively equivalent (it makes no difference how many times a line is rotated through a full circle because it always ends up in the same place). However, this is not always the case. For example, when tracing a curve such as a spiral using polar coordinates, an extra full turn gives rise to a quite different point on the curve.

Units

Angles are considered dimensionless, since they are defined as the ratio of lengths. There are, however, several units used to measure angles, depending on the choice of the constant k in the formula above.

With the notable exception of the radian, most units of angular measurement are defined such that one full circle (i.e. one revolution) is equal to n units, for some whole number n (for example, in the case of degrees, n = 360). This is equivalent to setting $k = n/2\pi$ in the formula above.

- The degree, denoted by a small superscript circle (°) is 1/360 of a full circle, so one full circle is 360°. One advantage of this old sexagesimal subunit is that many angles common in simple geometry are measured as a whole number of degrees. (The problem of having all "interesting" angles measured as whole numbers is of course insolvable.) Fractions of a degree may be written in normal decimal notation (e.g., 3.5° for three and a half degrees), but the following sexagesimal subunits of the "degree-minute-second" system are also in use, especially for geographical coordinates and in astronomy and ballistics:

  - The minute of arc (or MOA, arcminute, or just minute) is 1/60 of a degree. It is denoted by a single prime (′). For example, 3° 30′ is equal to 3 + 30/60 degrees, or 3.5 degrees. A mixed format with decimal fractions is also sometimes used, e.g., 3° 5.72′ = 3 + 5.72/60 degrees. A nautical mile was historically defined as a minute of arc along a great circle of the Earth.

  - The second of arc (or arcsecond, or just second) is 1/60 of a minute of arc and 1/3600 of a degree. It is denoted by a double prime (″). For example, 3° 7′ 30″ is equal to 3 + 7/60 + 30/3600 degrees, or 3.125 degrees.

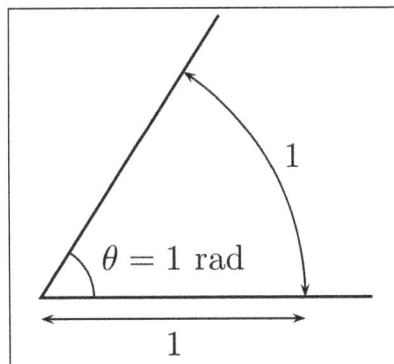

θ = s/r rad = 1 rad.

- The radian is the angle subtended by an arc of a circle that has the same length as the circle's radius ($k = 1$ in the formula given earlier). One full circle is 2π radians, and one radian is $180/\pi$ degrees, or about 57.2958 degrees. The radian is abbreviated rad, though this symbol is often omitted in mathematical texts, where radians are assumed unless specified otherwise. The radian is used in virtually all mathematical work beyond simple practical geometry, due, for example, to the pleasing and "natural" properties that the trigonometric functions display when their arguments are in radians. The radian is the (derived) unit of angular measurement in the SI system.

- The mil is approximately equal to a milliradian.

- The full circle (or revolution, rotation, full turn or cycle) is one complete revolution. The revolution and rotation are abbreviated rev and rot, respectively, but just r in rpm (revolutions per minute). 1 full circle = 360° = 2π rad = 400 gon = 4 right angles.

- The right angle is 1/4 of a full circle. It is the unit used in Euclid's Elements. 1 right angle = 90° = $\pi/2$ rad = 100 gon.

- The angle of the equilateral triangle is 1/6 of a full circle. It was the unit used by the Babylonians, and is especially easy to construct with ruler and compasses. The degree, minute of arc and second of arc are sexagesimal subunits of the Babylonian unit. One Babylonian unit = 60° = $\pi/3$ rad ≈ 1.047197551 rad.

- The grad, also called grade, gradian, or gon is 1/400 of a full circle, so one full circle is 400 grads and a right angle is 100 grads. It is a decimal subunit of the right angle. A kilometer was historically defined as a centi-gon of arc along a great circle of the Earth, so the kilometer is the decimal analog to the sexagesimal nautical mile. The gon is used mostly in triangulation.

- The point, used in navigation, is 1/32 of a full circle. It is a binary subunit of the full circle. Naming all 32 points on a compass rose is called "boxing the compass." 1 point = 1/8 of a right angle = 11.25° = 12.5 gon.

- The astronomical hour angle is 1/24 of a full circle. The sexagesimal subunits were called minute of time and second of time (even though they are units of angle). 1 hour = 15° = $\pi/12$ rad = 1/6 right angle ≈ 16.667 gon.

- The binary degree, also known as the binary radian (or brad), is 1/256 of a full circle. The binary degree is used in computing so that an angle can be efficiently represented in a single byte.

- The grade of a slope, or gradient, is not truly an angle measure (unless it is explicitly given in degrees, as is occasionally the case). Instead it is equal to the tangent of the angle, or sometimes the sine. Gradients are often expressed as a percentage. For the usual small values encountered (less than 5%), the grade of a slope is approximately the measure of an angle in radians.

Positive and Negative Angles

A convention universally adopted in mathematical writing is that angles given a sign are positive angles if measured counterclockwise, and negative angles if measured clockwise, from a given line. If no line is specified, it can be assumed to be the x-axis in the Cartesian plane. In many geometrical situations a negative angle of −θ is effectively equivalent to a positive angle of "one full rotation less θ." For example, a clockwise rotation of 45° (that is, an angle of −45°) is often effectively equivalent to a counterclockwise rotation of 360° − 45° (that is, an angle of 315°).

In three dimensional geometry, "clockwise" and "counterclockwise" have no absolute meaning, so the direction of positive and negative angles must be defined relative to some reference, which is typically a vector passing through the angle's vertex and perpendicular to the plane in which the rays of the angle lie.

In navigation, bearings are measured from north, increasing clockwise, so a bearing of 45 degrees is north-east. Negative bearings are not used in navigation, so north-west is 315 degrees.

TYPES OF ANGLES

There are many different types of angles. We will define them in this lesson. Study the images carefully so you understand them.

Acute Angle

An angle whose measure is less than 90 degrees. The following is an acute angle.

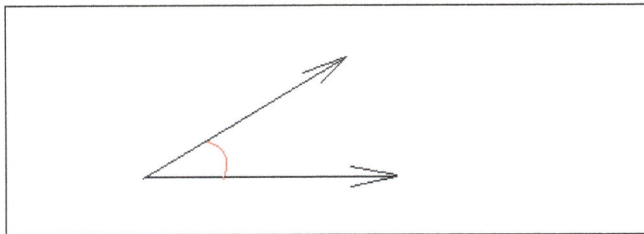

Acute angle.

Right Angle

An angle whose measure is 90 degrees. The following is a right angle.

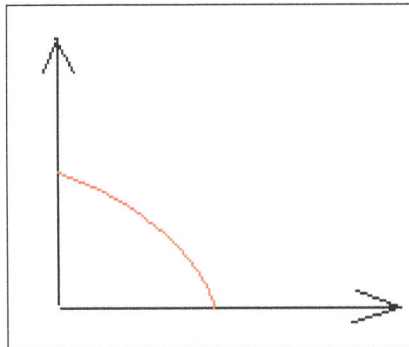

Right angle.

Obtuse Angle

An angle whose measure is bigger than 90 degrees but less than 180 degrees. Thus, it is between 90 degrees and 180 degrees. The following is an obtuse angle.

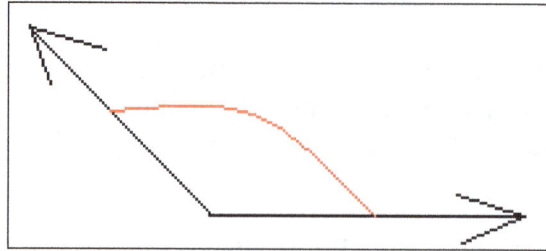

Obtuse angle.

Straight Angle

An angle whose measure is 180 degrees.Thus, a straight angle look like a straight line. The following is a straight angle.

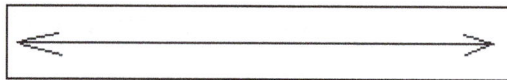

Straight angle.

Reflex Angle

An angle whose measure is bigger than 180 degrees but less than 360 degrees.The following is a reflex angle.

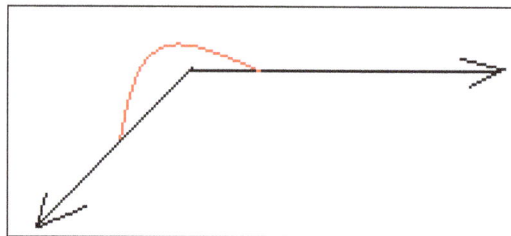

Reflex angle.

Adjacent Angles

Angle with a common vertex and one common side. <1 and <2, are adjacent angles.

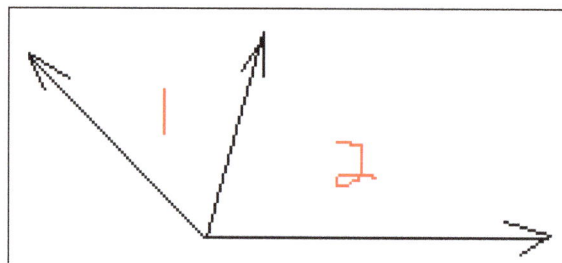

Adjacent angles.

Complementary Angles

Two angles whose measures add to 90 degrees. Angle 1 and angle 2 are complementary angles because together they form a right angle.

Note that angle 1 and angle 2 do not have to be adjacent to be complementary as long as they add up to 90 degrees.

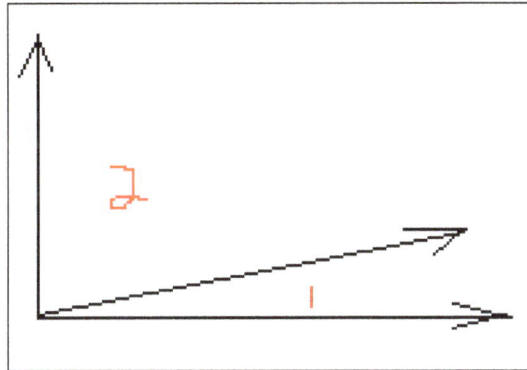

Complementary angles.

Supplementary Angles

Two angles whose measures add to 180 degrees. The following are supplementary angles.

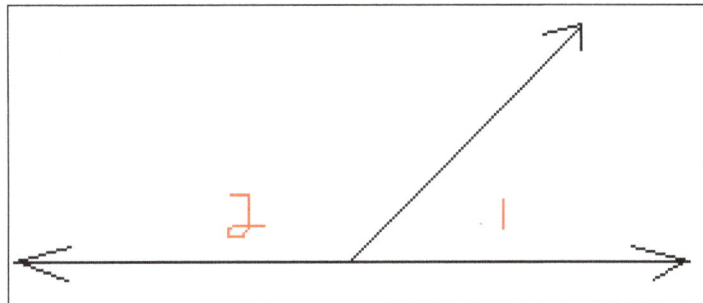

Supplementary angles.

Vertical Angles

Angles that have a common vertex and whose sides are formed by the same lines. The following(angle 1 and angle 2) are vertical angles.

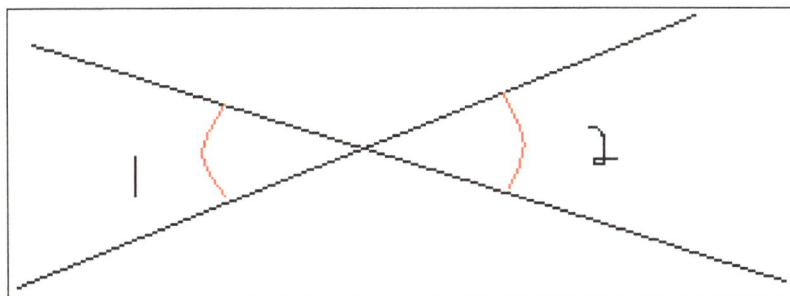

Vertical angles.

When two parallel lines are crossed by a third line(Transversal), 8 angles are formed. Take a look at the following figure:

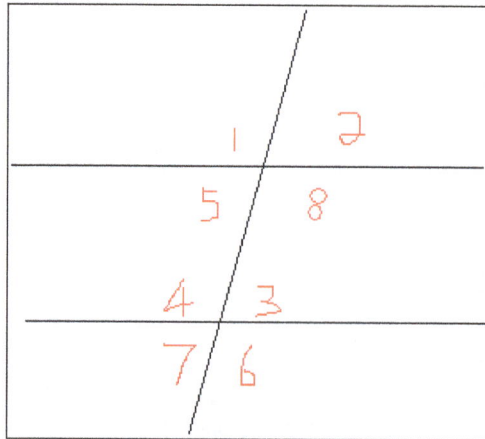

Transversal lines.

- Angles 3,4,5,8 are interior angles.

- Angles 1,2,6,7 are exterior angles.

Alternate Interior Angles

- Pairs of interior angles on opposite sides of the transversal.

- For instance, angle 3 and angle 5 are alternate interior angles. Angle 4 and angle 8 are also alternate interior angles.

Alternate Exterior Angles

- Pairs of exterior angles on opposite sides of the transversal.

- Angle 2 and angle 7 are alternate exterior angles.

Corresponding Angles

Pairs of angles that are in similar positions.

- Angle 3 and angle 2 are corresponding angles.

- Angle 5 and angle 7 are corresponding angles.

3

Geometrical Figures

There are a number of figures in geometry that are essential in the concepts and principles of mathematics. Scalene Triangle, Isosceles Triangle, Equilateral Triangle, Right Triangle, Square, Rhombus, Kite, Parallelogram, Trapezium, Sphere, Cone, Cuboid, etc. This chapter closely examines these figures of geometry to provide an extensive understanding of the subject.

TRIANGLE

A triangle is a 3-sided polygon sometimes (but not very commonly) called the trigon. Every triangle has three sides and three angles, some of which may be the same. The sides of a triangle are given special names in the case of a right triangle, with the side opposite the right angle being termed the hypotenuse and the other two sides being known as the legs. All triangles are convex and bicentric. That portion of the plane enclosed by the triangle is called the triangle interior, while the remainder is the exterior.

The study of triangles is sometimes known as triangle geometry, and is a rich area of geometry filled with beautiful results and unexpected connections. In 1816, while studying the Brocard points of a triangle, Crelle exclaimed, "It is indeed wonderful that so simple a figure as the triangle is so inexhaustible in properties. How many as yet unknown properties of other figures may there not be?"

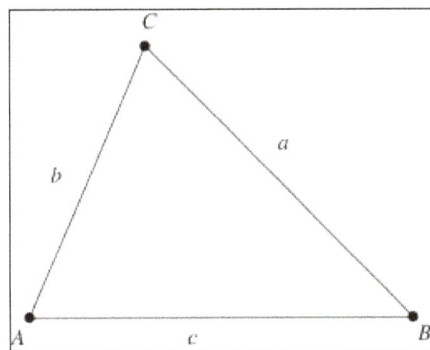
Triangle

It is common to label the vertices of a triangle in counterclockwise order as either A, B, C (or A_1, A_2, A_3). The vertex angles are then given the same symbols as the vertices themselves. The symbols α, β, γ (or $\alpha_1 \alpha_2 \alpha_3$) are also sometimes used, but this convention results in unnecessary confusion with the common notation for trilinear coordinates alpha:beta:gamma, and so is not recommend-ed. The sides opposite the angles A, B, and C (or A_1, A_2, A_3) are then labeled a, b, c (or a_1, a_2, a_3), with these symbols also indicating the lengths of the sides (just as the symbols at the vertices indi-cate the vertices themselves as well as the vertex angles, depending on context).

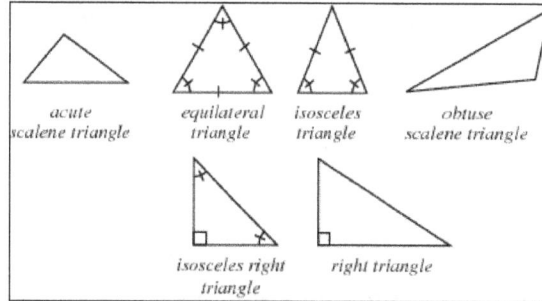

Triangles

A triangle is said to be acute if all three of its angles are all acute, a triangle having an obtuse angle is called an obtuse triangle, and a triangle with a right angle is called right. A triangle with all sides equal is called equilateral, a triangle with two sides equal is called isosceles, and a triangle with all sides a different length is called scalene. A triangle can be simultaneously right and isosceles, in which case it is known as an isosceles right triangle.

The semiperimeter s of a triangle is defined as half its perimeter:

$S = \frac{1}{2} p$

$= \frac{1}{2} (a+b+c).$

The area of a triangle can given by Heron's formula:

$$\Delta = \sqrt{(s(s-a)(s-b)(s-c)}.$$

There are also many other formulas for the triangle area.

The definition of the semiperimeter leads to the definitions:

$s_a = 1/2 (b+c-a)$

$= s - a$

$= r \cot (1/2\ A)$

$s_b = \frac{1}{2} (c + a - b)$

$= s - b$

$= r \cot (1/2\ B)$

$s_c = \frac{1}{2} (a + b - c)$

$= s - c$

$= r \cot (1/2\ C),$

where r is the inradius. A similar set of relations hold for Conway triangle notation S, S_A, S_B, and S_C.

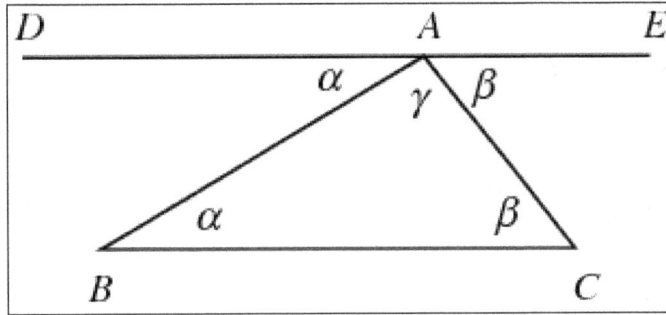

Triangle Angles.

The sum of angles in a triangle is 180° =pi radians (at least in Euclidean geometry; this statement does not hold in non-Euclidean geometry). This can be established as follows. Let DAE‖BC (DAE be parallel to BC) in the above diagram, then the angles alpha and beta satisfy alpha=∠DAB=∠ABC and beta=∠EAC=∠ACB, as indicated. Adding γ, it follows that,

$$\alpha + \beta + \gamma = 180^0,$$

since the sum of angles for the line segment must equal two right angles. Therefore, the sum of angles in the triangle is also 180°.

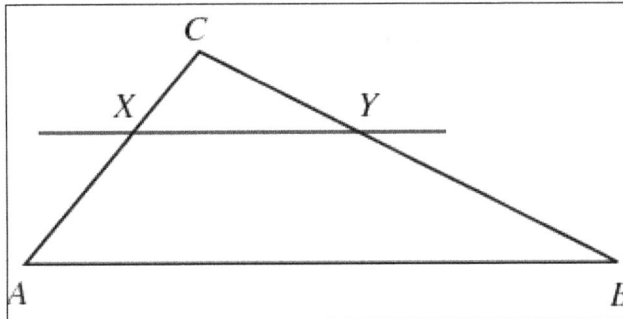

Triangle Parallel Line.

If a line is drawn parallel to one side of a triangle so that it intersects the other two sides, it divides them proportionally, i.e.,

$$\frac{(AX)}{(XC)} = \frac{(BY)}{(YC)}$$

In other words, a line parallel to a side of a triangle cutting the other two sides creates a triangle similar to the first.

Allowable side lengths a, b, and c for a triangle are given by the set of inequalities $a > 0, b > 0, c > 0$, and $a + b > c, b + c > a, a + c > b$, a statement that encapsulated in the so-called triangle inequality. The angles and sides of a triangle also satisfy an array of other beautiful triangle inequalities.

Specifying two angles A and B and a side a uniquely determines a triangle with area,

$$\Delta = \frac{a^2 \sin B \sin C}{2 \sin A}$$

$$= \frac{a^2 \sin B \sin (\pi - A - B)}{2 \sin A}$$

Specifying an angle A, a side c, and an angle B uniquely specifies a triangle with area:

$$\Delta = \frac{c^2}{2(\cot A + \cot B)}$$

Given a triangle with two sides, a the smaller and c the larger, and one known angle A, acute and opposite a, if $\sin A < a/c$, there are two possible triangles. If $\sin A = a/c$, there is one possible triangle. If $\sin A > a/c$, there are no possible triangles. This is the ASS theorem. Let a be the base length and h be the height. Then:

$$\Delta = \frac{1}{2} a h$$

$$= \frac{1}{2} a c \sin B$$

Finally, if all three sides are specified, a unique triangle is determined with area given by Heron's formula or by:

$$\Delta = \frac{abc}{4R},$$

where R is the circumradius. This is the SSS theorem.

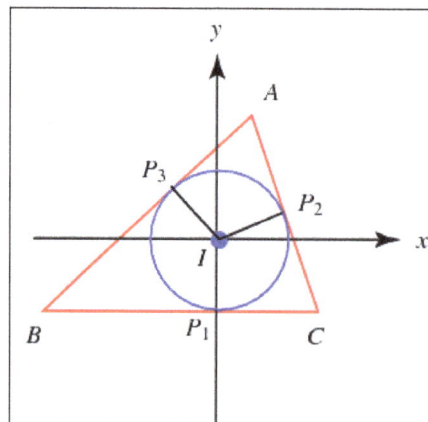

Trilinear Coordinates.

In triangle geometry, it is frequently very convenient to use a triple of coordinates defined relative to the distances from each side of a given so-called reference triangle. One form of such coordinates is known as trilinear coordinates $\alpha : \beta : \gamma$, with all coordinates having the same sign corresponding to the triangle interior, one coordinate zero corresponding to a point on a side, two coordinates zero corresponding to a vertex, and coordinates having different signs corresponding to the triangle exterior.

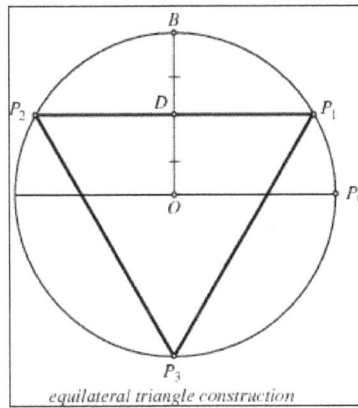

equilateral triangle construction

Triangle Construction.

The straightedge and compass construction of the triangle can be accomplished as follows. In the above figure, take OP_0 as a radius and draw $OB \perp OP_0$. Then bisect OB and construct $P_2P_1 \parallel OP_0$. Extending BO to locate P_3 then gives the equilateral triangle $\triangle P_1 P_2 P_3$. Another construction proceeds by drawing a circle of the desired radius r centered at a point   . Choose a point B on the circle's circumference and draw another circle of radius r centered at B. The two circles intersect at two points, P_1 and P_2, and P_3 is the second point at which the line B_0 intersects the first circle.

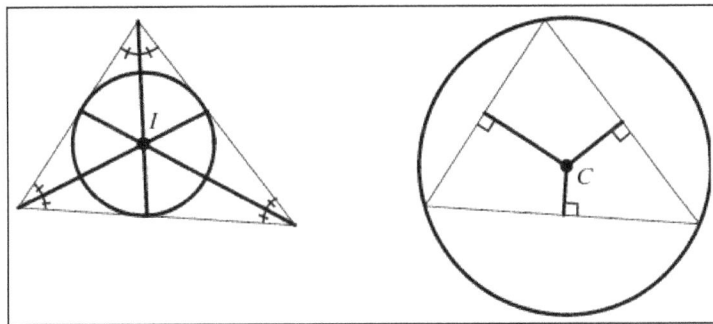

Incircle Circumcircle.

In Proposition IV.4 of the Elements, Euclid showed how to inscribe a circle (the incircle) in a given triangle by locating the incenter I as the point of intersection of angle bisectors. In Proposition IV.5, he showed how to circumscribe a circle (the circumcircle) about a given triangle by locating the circumcenter O as the point of intersection of the perpendicular bisectors. Unlike a general polygon with $n >= 4$ sides, a triangle always has both a circumcircle and an incircle. such polygons are called bicentric polygons.

A triangle with sides a, b, and c can be constructed by selecting vertices (0, 0), (c,0), and (x,y), then solving:

$$x^2 + y^2 = b^2$$
$$(x-c)^2 + y^2 = a^2$$

simultaneously to obtain,

$$x = \frac{-a^2 + b^2 + c^2}{2c}$$

$$= b \cos A$$

$$y = \pm \frac{\sqrt{(-a+b+c)(a-b+c)(a+b-c)(a+b+c)}}{2c}$$

$$= \pm \frac{2\Delta}{c}.$$

The angles of a triangle satisfy the law of cosines:

$$\cos A = \frac{b^2 + c^2 - a^2}{2bc},$$

as well as:

$$\cot A = \frac{b^2 + c^2 - a^2}{4\Delta}$$

where Delta is the area. The latter gives the pretty identity:

$$\cot A + \cot B + \cot C = \frac{(a^2 + b^2 + c^2)}{4\Delta}.$$

In addition,

$$\tan A + \tan B + \tan C = \tan A \tan B \tan C$$

and,

$$\cot B \cot C + \cot C \cot A + \cot A \cot B = 1$$
$$\tan A \cot B \cot C + \tan B \cot C \cot A + \tan C \cot A \cot B$$
$$= \tan A + \tan B + \tan C + 2(\cot A + \cot B + \cot C)$$

and,

$$\cot\left(\frac{1}{2}A\right) + \cot\left(\frac{1}{2}B\right) + \cot\left(\frac{1}{2}C\right)$$
$$= \cot\left(\frac{1}{2}A\right)\cot\left(\frac{1}{2}B\right)\cot\left(\frac{1}{2}C\right).$$

Additional formulas include:

$$\cos^2 A + \cos^2 B + \cos^2 C + 2\cos A \cos B \cos C = 1,$$

and,

$$\cos(nA) = \cos[n(B+C)]$$
$$\cos(nB) = \cos[n(A+C)]$$
$$\cos(nC) = \cos[n(A+B)]$$

for even n.

Trigonometric functions of half angles in a triangle can be expressed in terms of the triangle sides as:

$$\cos\left(\frac{1}{2}A\right) = \sqrt{\frac{s(s-a)}{bc}}$$

$$\sin\left(\frac{1}{2}A\right) = \sqrt{\frac{(s-b)(s-c)}{bc}}$$

$$\tan\left(\frac{1}{2}A\right) = \sqrt{\frac{(s-b)(s-c)}{s(s-a)}},$$

where s is the semiperimeter.

Let S stand for a triangle side and A for an angle, and let a set of S^s and A^s be concatenated such that adjacent letters correspond to adjacent sides and angles in a triangle. Triangles are uniquely determined by specifying three sides (SSS theorem), two angles and a side (AAS theorem), or two sides with an adjacent angle (SAS theorem). In each of these cases, the unknown three quantities (there are three sides and three angles total) can be uniquely determined. Other combinations of sides and angles do not uniquely determine a triangle: three angles specify a triangle only modulo a scale size (AAA theorem), and one angle and two sides not containing it may specify one, two, or no triangles (ASS theorem).

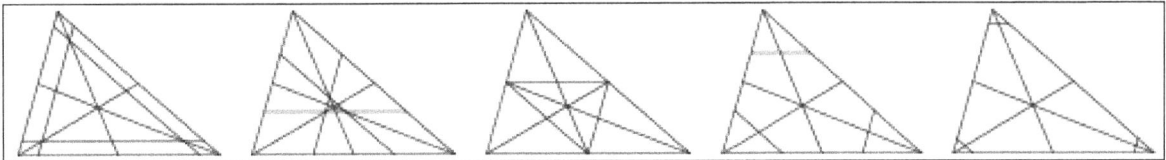
Side Parallels.

Dividing the sides of a triangle in a constant ratio $r < 1/2$ and then drawing lines parallel to the adjacent sides passing through each of these points gives line segments which intersect each other and one of the medians in three places. If $r > 1/2$, then the extensions of the side parallels intersect the extensions of the medians.

The medians bisect the area of a triangle, as do the side parallels with ratio $1+\sqrt{2}$. The envelope of the lines which bisect the area a triangle forms three hyperbolic arcs. The envelope is somewhat more complicated, however, for lines dividing the area of a triangle into a constant but unequal ratio.

SCALENE TRIANGLE

You have probably seen many triangles in your life. Perhaps you've even noticed that there are many different types of triangles. Some of these triangles have all three sides of the same length, some have two sides of the same length, and in some triangles, all three sides are different lengths.

Scalene triangles are triangles with three sides of different lengths. The math term for sides of different triangles is noncongruent sides, so you may also see this phrase in your math book. For example, a triangle with side lengths of 2 cm, 3 cm, and 4 cm would be a scalene triangle. A triangle with side lengths of 2 cm, 2 cm, and 3 cm would not be scalene, since two of the sides have the same length.

The hash marks on each side of the scalene triangles show that all sides are different lengths.

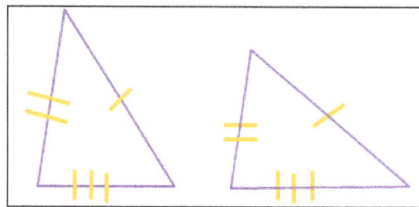

Two Scalene Triangles.

Properties of Scalene Triangles

The most important property of scalene triangles is that they have three sides of different lengths. However, they have some other important properties, too. Like other triangles, all the angles inside a scalene triangle add up to 180°. And just like all the sides of a scalene triangle have different lengths, all the angles of a scalene triangle have different measures.

Let's take a look at some examples of triangles that we can classify as scalene or not scalene by their angle measures:

- 40 degrees - 50 degrees - 90 degrees is a scalene triangle since all the angle measures are different.

- 60 degrees - 60 degrees - 60 degrees is not a scalene triangle since the angle measures are not all different.

- 120 degrees - 10 degrees - 50 degrees is a scalene triangle since all the angle measures are different.

The triangle on the left is scalene because it has three different angles. The triangle on the right is NOT scalene because it has two angles of the same size.

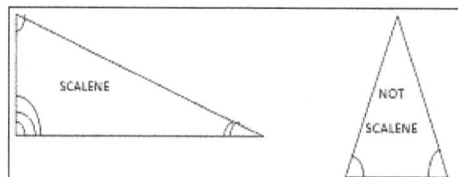

Scalene and not Scalene.

In addition, there are some other properties that you might find useful as you encounter scalene triangles in math problems.

- The longest side of the triangle is opposite the largest angle. This means that in the 120-10-50 triangle above, the longest side of the triangle is across from the 120-degree angle.

- The shortest side of the triangle is opposite the smallest angle. This means that in the 120-10-50 triangle, the shortest side is located across from the 10-degree angle.

- This picture shows these concepts more clearly. In the picture, the bright green side, labeled longest side, is across from the largest angle, B. The shortest side, in black, is across from the smallest angle, angle C.

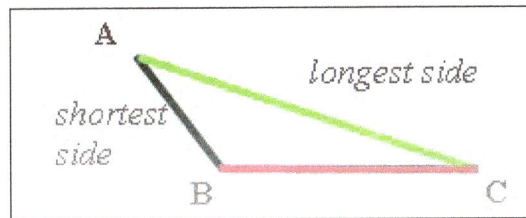

ISOSCELES TRIANGLE

In geometry, an isosceles triangle is a triangle that has two sides of equal length. Sometimes it is specified as having *exactly* two sides of equal length, and sometimes as having *at least* two sides of equal length, the latter version thus including the equilateral triangle as a special case. Examples of isosceles triangles include the isosceles right triangle, the golden triangle, and the faces of bipyramids and certain Catalan solids.

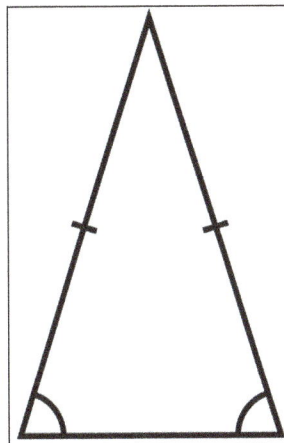

Isosceles triangle with vertical axis of symmetry.

The mathematical study of isosceles triangles dates back to ancient Egyptian mathematics and Babylonian mathematics. Isosceles triangles have been used as decoration from even earlier times, and appear frequently in architecture and design, for instance in the pediments and gables of buildings.

The two equal sides are called the legs and the third side is called the base of the triangle. The other dimensions of the triangle, such as its height, area, and perimeter, can be calculated by simple formulas from the lengths of the legs and base. Every isosceles triangle has an axis of symmetry along the perpendicular bisector of its base. The two angles opposite the legs are equal and are always acute, so the classification of the triangle as acute, right, or obtuse depends only on the angle between its two legs.

Terminology and Classification

Euclid defined an isosceles triangle as a triangle with exactly two equal sides, but modern treatments prefer to define isosceles triangles as having at least two equal sides. The difference between these two definitions is that the modern version makes equilateral triangles (with three equal sides) a special case of isosceles triangles. A triangle that is not isosceles (having three unequal sides) is called scalene. "Isosceles" is a compound word, made from the Greek roots "isos" (equal) and "skelos" (leg). The same word is used, for instance, for isosceles trapezoids, trapezoids with two equal sides, and for isosceles sets, sets of points every three of which form an isosceles triangle.

In an isosceles triangle that has exactly two equal sides, the equal sides are called legs and the third side is called the base. The angle included by the legs is called the *vertex angle* and the angles that have the base as one of their sides are called the *base angles*. The vertex opposite the base is called the apex. In the equilateral triangle case, since all sides are equal, any side can be called the base.

Special Isosceles Triangles

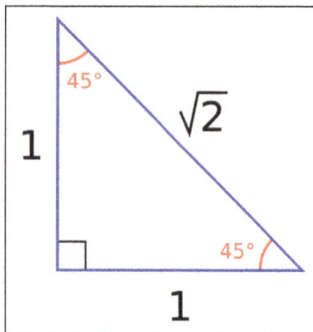

Isosceles right triangle.

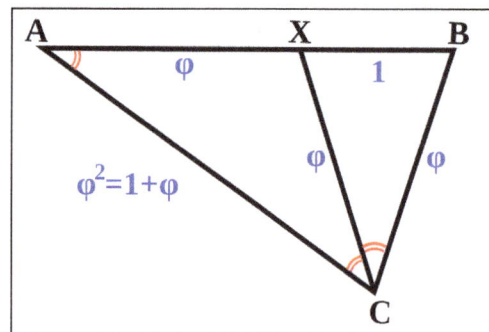

A golden triangle subdivided into a smaller golden triangle and golden gnomon.

Whether an isosceles triangle is acute, right or obtuse depends only on the angle at its apex. In Euclidean geometry, the base angles can not be obtuse (greater than 90°) or right (equal to 90°) because their measures would sum to at least 180°, the total of all angles in any Euclidean triangle. Since a triangle is obtuse or right if and only if one of its angles is obtuse or right, respectively, an isosceles triangle is obtuse, right or acute if and only if its apex angle is respectively obtuse, right or acute. In Edwin Abbott's book *Flatland*, this classification of shapes was used as a satire of social hierarchy: isosceles triangles represented the working class, with acute isosceles triangles higher in the hierarchy than right or obtuse isosceles triangles.

As well as the isosceles right triangle, several other specific shapes of isosceles triangles have been studied. These include the Calabi triangle (a triangle with three congruent inscribed squares), the golden triangle and golden gnomon (two isosceles triangles whose sides and base are in the golden ratio), the 80-80-20 triangle appearing in the Langley's Adventitious Angles puzzle, and the 30-30-120 triangle of the triakis triangular tiling. Five Catalan solids, the triakis tetrahedron, triakis octahedron, tetrakis hexahedron, pentakis dodecahedron, and triakis icosahedron, each have isosceles-triangle faces, as do infinitely many pyramids and bipyramids.

Formulas

Height

For any isosceles triangle, the following six line segments coincide:

- The altitude, a line segment from the apex perpendicular to the base.

- The angle bisector from the apex to the base.

- The median from the apex to the midpoint of the base.

- The perpendicular bisector of the base within the triangle.

- The segment within the triangle of the unique axis of symmetry of the triangle.

- The segment within the triangle of the Euler line of the triangle.

Their common length is the height h of the triangle. If the triangle has equal sides of length a and base of length b, the general triangle formulas for the lengths of these segments all simplify to:

$$h = \frac{1}{2}\sqrt{4a^2 - b^2}.$$

This formula can also be derived from the Pythagorean theorem using the fact that the altitude bisects the base and partitions the isosceles triangle into two congruent right triangles.

The Euler line of any triangle goes through the triangle's orthocenter (the intersection of its three altitudes), its centroid (the intersection of its three medians), and its circumcenter (the intersection of the perpendicular bisectors of its three sides, which is also the center of the circumcircle that passes through the three vertices). In an isosceles triangle with exactly two equal sides, these three points are distinct, and (by symmetry) all lie on the symmetry axis of the triangle, from which it follows that the Euler line coincides with the axis of symmetry. The incenter of the triangle also lies on the Euler line, something that is not true for other triangles. If any two of an angle bisector, median, or altitude coincide in a given triangle, that triangle must be isosceles.

Area

The area T of an isosceles triangle can be derived from the formula for its height, and from the general formula for the area of a triangle as half the product of base and height:

$$T = \frac{b}{4}\sqrt{4a^2 - b^2}.$$

The same area formula can also be derived from Heron's formula for the area of a triangle from its three sides. However, applying Heron's formula directly can be numerically unstable for isosceles triangles with very sharp angles, because of the near-cancellation between the semiperimeter and side length in those triangles.

If the apex angle (θ) and leg lengths (a) of an isosceles triangle are known, then the area of that triangle is:

$$T = \frac{1}{2}a^2 \sin\theta.$$

This is a special case of the general formula for the area of a triangle as half the product of two sides times the sine of the included angle.

Perimeter

The perimeter p of an isosceles triangle with equal sides a and base b is just:

$$p = 2a + b.$$

As in any triangle, the area T and perimeter p are related by the isoperimetric inequality:

$$p^2 > 12\sqrt{3}T.$$

This is a strict inequality for isosceles triangles with sides unequal to the base, and becomes an equality for the equilateral triangle. The area, perimeter, and base can also be related to each other by the equation:

$$2pb^3 - p^2b^2 + 16T^2 = 0.$$

If the base and perimeter are fixed, then this formula determines the area of the resulting isosceles triangle, which is the maximum possible among all triangles with the same base and perimeter On the other hand, if the area and perimeter are fixed, this formula can be used to recover the base length, but not uniquely: there are in general two distinct isosceles triangles with given area T and perimeter p. When the isoperimetric inequality becomes an equality, there is only one such triangle, which is equilateral.

Angle Bisector Length

If the two equal sides have length a and the other side has length b, then the internal angle bisector t from one of the two equal-angled vertices satisfies:

$$\frac{2ab}{a+b} > t > \frac{ab\sqrt{2}}{a+b}$$

as well as:

$$t < \frac{4a}{3};$$

and conversely, if the latter condition holds, an isosceles triangle parametrized by a and t exists.

The Steiner–Lehmus theorem states that every triangle with two angle bisectors of equal lengths is isosceles. It was formulated in 1840 by C. L. Lehmus. Its other namesake, Jakob Steiner, was one of the first to provide a solution. Although originally formulated only for internal angle bisectors, it works for many (but not all) cases when, instead, two external angle bisectors are equal. The 30-30-120 isosceles triangle makes a boundary case for this variation of the theorem, as it has four equal angle bisectors (two internal, two external).

Radii

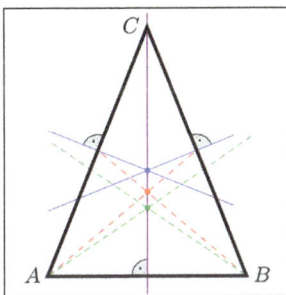

Isosceles triangle showing its circumcenter (blue), centroid (red), incenter (green), and symmetry axis (purple).

The inradius and circumradius formulas for an isosceles triangle may be derived from their formulas for arbitrary triangles. The radius of the inscribed circle of an isosceles triangle with side length a, base b, and height h is:

$$\frac{2ab - b^2}{4h}.$$

The center of the circle lies on the symmetry axis of the triangle, this distance above the base. An isosceles triangle has the largest possible inscribed circle among the triangles with the same base and apex angle, as well as also having the largest area and perimeter among the same class of triangles.

The radius of the circumscribed circle is:

$$\frac{a^2}{2h}.$$

The center of the circle lies on the symmetry axis of the triangle, this distance below the apex.

Inscribed Square

For any isosceles triangle, there is a unique square with one side collinear with the base of the triangle and the opposite two corners on its sides. The Calabi triangle is a special isosceles triangle with the property that the other two inscribed squares, with sides collinear with the sides of the triangle, are of the same size as the base square. A much older theorem, preserved in the works of Hero of Alexandria, states that, for an isosceles triangle with base b and height h, the side length of the inscribed square on the base of the triangle is:

$$\frac{bh}{b+h}.$$

Isosceles Subdivision of other Shapes

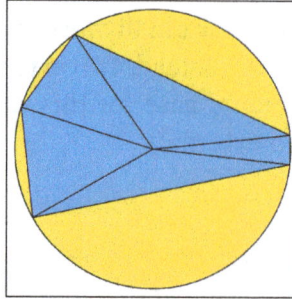

Partition of a cyclic pentagon into isosceles triangles by radii of its circumcircle.

For any integer $n \geq 4$, any triangle can be partitioned into n isosceles triangles. In a right triangle, the median from the hypotenuse (that is, the line segment from the midpoint of the hypotenuse to the right-angled vertex) divides the right triangle into two isosceles triangles. This is because the midpoint of the hypotenuse is the center of the circumcircle of the right triangle, and each of the two triangles created by the partition has two equal radii as two of its sides. Similarly, an acute triangle can be partitioned into three isosceles triangles by segments from its circumcenter, but this method does not work for obtuse triangles, because the circumcenter lies outside the triangle.

Generalizing the partition of an acute triangle, any cyclic polygon that contains the center of its circumscribed circle can be partitioned into isosceles triangles by the radii of this circle through its vertices. The fact that all radii of a circle have equal length implies that all of these triangles are isosceles. This partition can be used to derive a formula for the area of the polygon as a function of its side lengths, even for cyclic polygons that do not contain their circumcenters. This formula generalizes Heron's formula for triangles and Brahmagupta's formula for cyclic quadrilaterals.

Either diagonal of a rhombus divides it into two congruent isosceles triangles. Similarly, one of the two diagonals of a kite divides it into two isosceles triangles, which are not congruent except when the kite is a rhombus.

Applications

In Architecture and Design

Obtuse isosceles pediment of the Pantheon, Rome.

Isosceles triangles commonly appear in architecture as the shapes of gables and pediments. In ancient Greek architecture and its later imitations, the obtuse isosceles triangle was used; in Gothic architecture this was replaced by the acute isosceles triangle.

Acute isosceles gable over the Saint-Etienne portal, Notre-Dame de Paris.

In the architecture of the Middle Ages, another isosceles triangle shape became popular: the Egyptian isosceles triangle. This is an isosceles triangle that is acute, but less so than the equilateral triangle; its height is proportional to 5/8 of its base. The Egyptian isosceles triangle was brought back into use in modern architecture by Dutch architect Hendrik Petrus Berlage.

Detailed view of a modified Warren truss with verticals.

Warren truss structures, such as bridges, are commonly arranged in isosceles triangles, although sometimes vertical beams are also included for additional strength. Surfaces tessellated by obtuse isosceles triangles can be used to form deployable structures that have two stable states: an unfolded state in which the surface expands to a cylindrical column, and a folded state in which it folds into a more compact prism shape that can be more easily transported.

In graphic design and the decorative arts, isosceles triangles have been a frequent design element in cultures around the world from at least the Early Neolithic to modern times. They are a common design element in flags and heraldry, appearing prominently with a vertical base, for instance, in the flag of Guyana, or with a horizontal base in the flag of Saint Lucia, where they form a stylized image of a mountain island.

They also have been used in designs with religious or mystic significance, for instance in the Sri Yantra of Hindu meditational practice.

In other Areas of Mathematics

If a cubic equation with real coefficients has three roots that are not all real numbers, then when these roots are plotted in the complex plane as an Argand diagram they form vertices of an isosceles triangle whose axis of symmetry coincides with the horizontal (real) axis. This is because the complex roots are complex conjugates and hence are symmetric about the real axis.

In celestial mechanics, the three-body problem has been studied in the special case that the three bodies form an isosceles triangle, because assuming that the bodies are arranged in this way reduces the number of degrees of freedom of the system without reducing it to the solved Lagrangian point case when the bodies form an equilateral triangle. The first instances of the three-body problem shown to have unbounded oscillations were in the isosceles three-body problem.

EQUILATERAL TRIANGLE

In geometry, an equilateral triangle is a triangle in which all three sides are equal. In the familiar Euclidean geometry, an equilateral triangle is also equiangular; that is, all three internal angles are also congruent to each other and are each 60°. It is also a regular polygon, so it is also referred to as a regular triangle.

Principal Properties

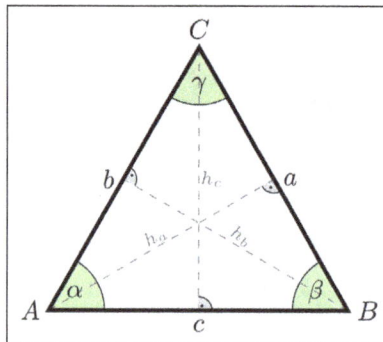

An equilateral triangle. It has equal sides $(a = b = c)$, equal angles $(\alpha = \beta = \gamma)$, and equal altitudes $(h_a = h_b = h_c)$.

Denoting the common length of the sides of the equilateral triangle as a, we can determine using the Pythagorean theorem that:

- The area is $A = \dfrac{\sqrt{3}}{4} a^2$.

- The perimeter is $p = 3a$.

- The radius of the circumscribed circle is $R = \dfrac{a}{\sqrt{3}}$.

- The radius of the inscribed circle is $r = \dfrac{\sqrt{3}}{6} a$ or $r = \dfrac{R}{2}$.

- The geometric center of the triangle is the center of the circumscribed and inscribed circles.

- The altitude (height) from any side is $h = \dfrac{\sqrt{3}}{2} a$.

Denoting the radius of the circumscribed circle as R, we can determine using trigonometry that:

- The area of the triangle is $A = \dfrac{3\sqrt{3}}{4} R^2$.

Many of these quantities have simple relationships to the altitude ("h") of each vertex from the opposite side:

- The area is $A = \dfrac{h^2}{\sqrt{3}}$.

- The height of the center from each side, or apothem, is $R = \dfrac{2h}{3}$.

- The radius of the circle circumscribing the three vertices is $R = \dfrac{2h}{3}$.

- The radius of the inscribed circle is $r = \dfrac{h}{3}$.

In an equilateral triangle, the altitudes, the angle bisectors, the perpendicular bisectors, and the medians to each side coincide.

Characterizations

A triangle ABC that has the sides a, b, c, semiperimeter s, area T, exradii r_a, r_b, r_c (tangent to a, b, c respectively), and where R and r are the radii of the circumcircle and incircle respectively, is equilateral if and only if any one of the statements in the following nine categories is true. Thus these are properties that are unique to equilateral triangles, and knowing that any one of them is true directly implies that we have an equilateral triangle.

Sides

- $a = b = c$

- $\dfrac{1}{a} + \dfrac{1}{b} + \dfrac{1}{c} = \dfrac{\sqrt{25Rr - 2r^2}}{4Rr}$

Semiperimeter

- $s = 2R + (3\sqrt{3} - 4)r$ (Blundon)

- $s^2 = 3r^2 + 12Rr$

- $s^2 = 3\sqrt{3}T$

- $s = 3\sqrt{3}r$

- $s = \dfrac{3\sqrt{3}}{2}R$

Angles

- $A = B = C = 60°$

- $\cos A + \cos B + \cos C = \dfrac{3}{2}$

- $\sin \dfrac{A}{2} \sin \dfrac{B}{2} \sin \dfrac{C}{2} = \dfrac{1}{8}$

Area

- $T = \dfrac{a^2 + b^2 + c^2}{4\sqrt{3}}$ (Weitzenböck)

- $T = \dfrac{\sqrt{3}}{4}(abc)^{\frac{2}{3}}$

Circumradius, Inradius and Exradii

- $R = 2r$ (Chapple-Euler)

- $9R^2 = a^2 + b^2 + c^2$

- $r = \dfrac{r_a + r_b + r_c}{9}$

- $r_a = r_b = r_c$

Equal Cevians

Three kinds of cevians coincide, and are equal, for (and only for) equilateral triangles:

- The three altitudes have equal lengths.
- The three medians have equal lengths.
- The three angle bisectors have equal lengths.

Coincident Triangle Centers

Every triangle center of an equilateral triangle coincides with its centroid, which implies that the equilateral triangle is the only triangle with no Euler line connecting some of the centers. For some pairs of triangle centers, the fact that they coincide is enough to ensure that the triangle is equilateral. In particular:

- A triangle is equilateral if any two of the circumcenter, incenter, centroid, or orthocenter coincide.

- It is also equilateral if its circumcenter coincides with the Nagel point, or if its incenter coincides with its nine-point center.

Six Triangles Formed by Partitioning by the Medians

For any triangle, the three medians partition the triangle into six smaller triangles.

- A triangle is equilateral if and only if any three of the smaller triangles have either the same perimeter or the same inradius.

- A triangle is equilateral if and only if the circumcenters of any three of the smaller triangles have the same distance from the centroid.

Points in the Plane

- A triangle is equilateral if and only if, for *every* point P in the plane, with distances p, q, and r to the triangle's sides and distances x, y, and z to its vertices,

$$4(p^2 + q^2 + r^2) \geq x^2 + y^2 + z^2.$$

Notable Theorems

Morley's trisector theorem states that, in any triangle, the three points of intersection of the adjacent angle trisectors form an equilateral triangle.

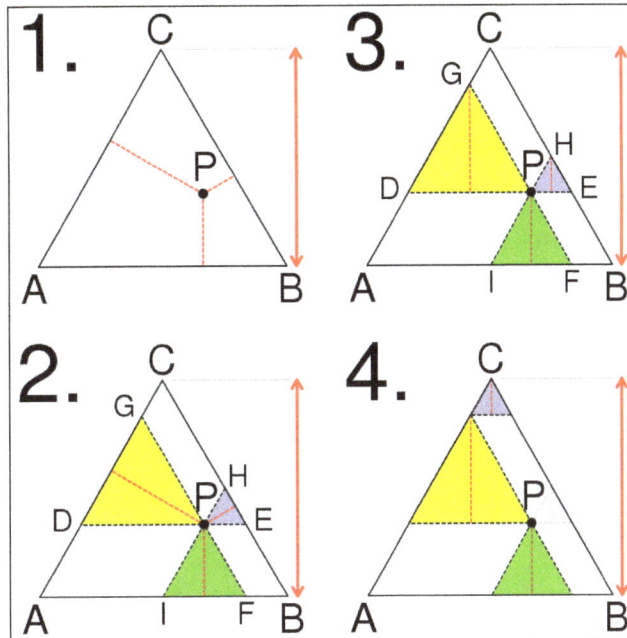

Visual proof of Viviani's theorem:

- Nearest distances from point P to sides of equilateral triangle ABC are shown.

- Lines DE, FG, and HI parallel to AB, BC and CA, respectively, define smaller triangles PHE, PFI and PDG.

- As these triangles are equilateral, their altitudes can be rotated to be vertical.

- As PGCH is a parallelogram, triangle PHE can be slid up to show that the altitudes sum to that of triangle ABC.

Napoleon's theorem states that, if equilateral triangles are constructed on the sides of any triangle, either all outward, or all inward, the centers of those equilateral triangles themselves form an equilateral triangle.

A version of the isoperimetric inequality for triangles states that the triangle of greatest area among all those with a given perimeter is equilateral.

Viviani's theorem states that, for any interior point P in an equilateral triangle with distances d, e, and f from the sides and altitude h,

$$d + e + f = h,$$

independent of the location of P.

Pompeiu's theorem states that, if P is an arbitrary point in the plane of an equilateral triangle ABC but not on its circumcircle, then there exists a triangle with sides of lengths PA, PB, and PC. That is, PA, PB, and PC satisfy the triangle inequality that the sum of any two of them is greater than the third. If P is on the circumcircle then the sum of the two smaller ones equals the longest and the triangle has degenerated into a line, this case is known as Van Schooten's theorem.

Other Properties

By Euler's inequality, the equilateral triangle has the smallest ratio R/r of the circumradius to the inradius of any triangle: specifically, $R/r = 2$.

The triangle of largest area of all those inscribed in a given circle is equilateral; and the triangle of smallest area of all those circumscribed around a given circle is equilateral.

The ratio of the area of the incircle to the area of an equilateral triangle, $\dfrac{\pi}{3\sqrt{3}}$, is larger than that of any non-equilateral triangle.

The ratio of the area to the square of the perimeter of an equilateral triangle, $\dfrac{1}{12\sqrt{3}}$ is larger than that for any other triangle.

If a segment splits an equilateral triangle into two regions with equal perimeters and with areas A_1 and A_2, then,

$$\frac{7}{9} \le \frac{A_1}{A_2} \le \frac{9}{7}.$$

If a triangle is placed in the complex plane with complex vertices z_1, z_2, and z_3, then for either non-real cube root ω of 1 the triangle is equilateral if and only if:

$$z_1 + \omega z_2 + \omega^2 z_3 = 0.$$

Given a point P in the interior of an equilateral triangle, the ratio of the sum of its distances from the vertices to the sum of its distances from the sides is greater than or equal to 2, equality holding when P is the centroid. In no other triangle is there a point for which this ratio is as small as 2. This is the Erdős–Mordell inequality; a stronger variant of it is Barrow's inequality, which replaces the perpendicular distances to the sides with the distances from P to the points where the angle bisectors of $\angle APB$, $\angle BPC$, and $\angle CPA$ cross the sides (A, B, and C being the vertices).

For any point P in the plane, with distances p, q, and t from the vertices A, B, and C respectively,

$$3(p^4 + q^4 + t^4 + a^4) = (p^2 + q^2 + t^2 + a^2)^2.$$

For any point P on the inscribed circle of an equilateral triangle, with distances p, q, and t from the vertices,

$$4(p^2 + q^2 + t^2) = 5a^2$$

and,

$$16(p^4 + q^4 + t^4) = 11a^4.$$

For any point P on the minor arc BC of the circumcircle, with distances p, q, and t from A, B, and C respectively,

$$p = q + t$$

and,

$$q^2 + qt + t^2 = a^2;$$

moreover, if point D on side BC divides PA into segments PD and DA with DA having length z and PD having length y, then,

$$z = \frac{t^2 + tq + q^2}{t + q},$$

which also equals $\frac{t^3 - q^3}{t^2 - q^2}$ if $t \neq q$,

$$\frac{1}{q} + \frac{1}{t} = \frac{1}{y},$$

which is the optic equation.

There are numerous triangle inequalities that hold with equality if and only if the triangle is equilateral.

An equilateral triangle is the most symmetrical triangle, having 3 lines of reflection and rotational symmetry of order 3 about its center. Its symmetry group is the dihedral group of order 6 D_3.

Equilateral triangles are the only triangles whose Steiner inellipse is a circle (specifically, it is the incircle).

The integer-sided equilateral triangle is the only triangle with integer sides and three rational angles as measured in degrees.

The equilateral triangle is the only acute triangle that is similar to its orthic triangle (with vertices at the feet of the altitudes) (the heptagonal triangle being the only obtuse one).

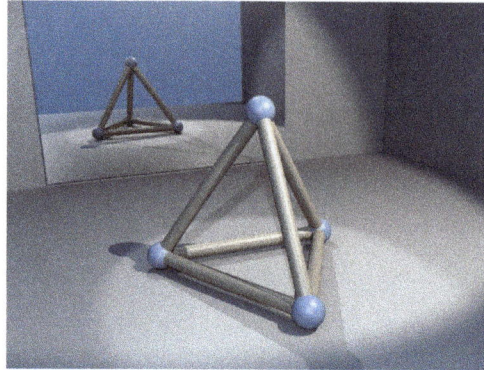

A regular tetrahedron is made of four equilateral triangles.

Equilateral triangles are found in many other geometric constructs. The intersection of circles whose centers are a radius width apart is a pair of equilateral arches, each of which can be inscribed with an equilateral triangle. They form faces of regular and uniform polyhedra. Three of the five Platonic solids are composed of equilateral triangles. In particular, the regular tetrahedron has four equilateral triangles for faces and can be considered the three-dimensional analogue of the shape. The plane can be tiled using equilateral triangles giving the triangular tiling.

Geometric Construction

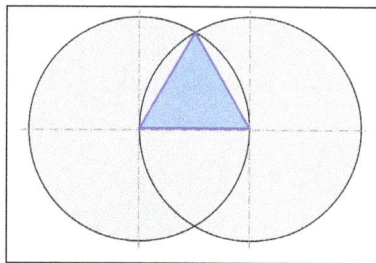

Construction of equilateral triangle with compass and straightedge.

An equilateral triangle is easily constructed using a straightedge and compass, because 3 is a Fermat prime. Draw a straight line, and place the point of the compass on one end of the line, and swing an arc from that point to the other point of the line segment. Repeat with the other side of the line. Finally, connect the point where the two arcs intersect with each end of the line segment.

An alternative method is to draw a circle with radius r, place the point of the compass on the circle and draw another circle with the same radius. The two circles will intersect in two points. An equilateral triangle can be constructed by taking the two centers of the circles and either of the points of intersection.

In both methods a by-product is the formation of vesica piscis.

The proof that the resulting figure is an equilateral triangle is the first proposition in Book I of Euclid's *Elements*.

Derivation of Area Formula

The area formula $A = \dfrac{\sqrt{3}}{4} a^2$ in terms of side length a can be derived directly using the Pythagorean theorem or using trigonometry.

Using the Pythagorean Theorem

The area of a triangle is half of one side a times the height h from that side:

$$A = \frac{1}{2} ah.$$

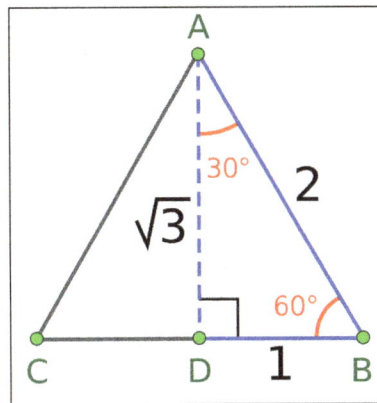

An equilateral triangle with a side of 2 has a height of √3 as the sine of 60° is √3/2.

The legs of either right triangle formed by an altitude of the equilateral triangle are half of the base a, and the hypotenuse is the side a of the equilateral triangle. The height of an equilateral triangle can be found using the Pythagorean theorem:

$$\left(\frac{a}{2}\right)^2 + h^2 = a^2$$

so that,

$$h = \frac{\sqrt{3}}{2} a.$$

Substituting h into the area formula $(1/2)ah$ gives the area formula for the equilateral triangle:

$$A = \frac{\sqrt{3}}{4} a^2.$$

Using Trigonometry

Using trigonometry, the area of a triangle with any two sides a and b, and an angle C between them is:

$$A = \frac{1}{2}ab\sin C.$$

Each angle of an equilateral triangle is 60°, so:

$$A = \frac{1}{2}ab\sin 60^{\circ}.$$

The sine of 60° is $\frac{\sqrt{3}}{2}$. Thus,

$$A = \frac{1}{2}ab \times \frac{\sqrt{3}}{2} = \frac{\sqrt{3}}{4}ab = \frac{\sqrt{3}}{4}a^2$$

since all sides of an equilateral triangle are equal.

ACUTE AND OBTUSE TRIANGLES

An acute triangle (or acute-angled triangle) is a triangle with three acute angles (less than 90°). An obtuse triangle (or obtuse-angled triangle) is a triangle with one obtuse angle (greater than 90°) and two acute angles. Since a triangle's angles must sum to 180° in Euclidean geometry, no Euclidean triangle can have more than one obtuse angle.

Acute and obtuse triangles are the two different types of oblique triangles — triangles that are not right triangles because they have no 90° angle.

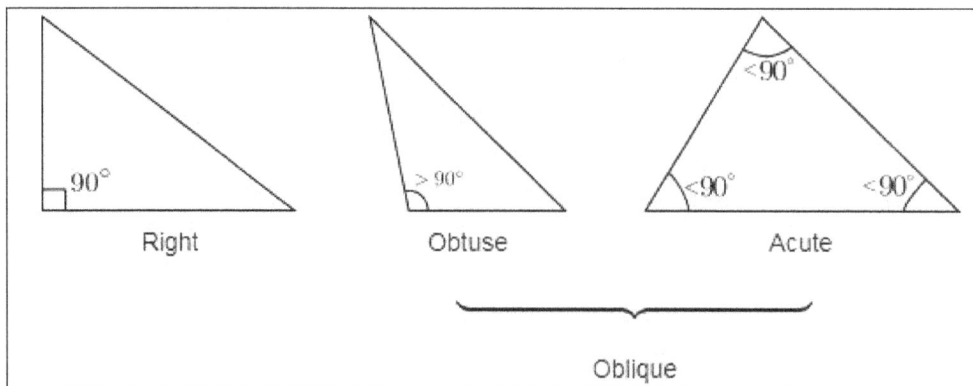

Properties

In all triangles, the centroid—the intersection of the medians, each of which connects a vertex with the midpoint of the opposite side—and the incenter—the center of the circle that is internally tangent to all three sides—are in the interior of the triangle. However, while the orthocenter and the circumcenter are in an acute triangle's interior, they are exterior to an obtuse triangle.

The orthocenter is the intersection point of the triangle's three altitudes, each of which perpendicularly connects a side to the opposite vertex. In the case of an acute triangle, all three of these segments lie entirely in the triangle's interior, and so they intersect in the interior. But for an obtuse triangle, the altitudes from the two acute angles intersect only the extensions of the opposite sides. These altitudes fall entirely outside the triangle, resulting in their intersection with each other (and hence with the extended altitude from the obtuse-angled vertex) occurring in the triangle's exterior.

Likewise, a triangle's circumcenter—the intersection of the three sides' perpendicular bisectors, which is the center of the circle that passes through all three vertices—falls inside an acute triangle but outside an obtuse triangle.

The right triangle is the in-between case: both its circumcenter and its orthocenter lie on its boundary.

In any triangle, any two angle measures A and B opposite sides a and b respectively are related according to:

$$A > B \quad \text{if and only if} \quad a > b.$$

This implies that the longest side in an obtuse triangle is the one opposite the obtuse-angled vertex.

An acute triangle has three inscribed squares, each with one side coinciding with part of a side of the triangle and with the square's other two vertices on the remaining two sides of the triangle. (In a right triangle two of these are merged into the same square, so there are only two distinct inscribed squares.) However, an obtuse triangle has only one inscribed square, one of whose sides coincides with part of the longest side of the triangle.

All triangles in which the Euler line is parallel to one side are acute. This property holds for side BC if and only if $(\tan B)(\tan C) = 3$.

Inequalities

Sides

If angle C is obtuse then for sides a, b, and c we have:

$$\frac{c^2}{2} < a^2 + b^2 < c^2,$$

with the left inequality approaching equality in the limit only as the apex angle of an isosceles triangle approaches 180°, and with the right inequality approaching equality only as the obtuse angle approaches 90°.

If the triangle is acute then:

$$a^2 + b^2 > c^2, \quad b^2 + c^2 > a^2, \quad c^2 + a^2 > b^2.$$

Altitude

If C is the greatest angle and h_c is the altitude from vertex C, then for an acute triangle:

$$\frac{1}{h_c^2} < \frac{1}{a^2} + \frac{1}{b^2},$$

with the opposite inequality if C is obtuse.

Medians

With longest side c and medians m_a and m_b from the other sides,

$$4c^2 + 9a^2b^2 > 16m_a^2 m_b^2$$

for an acute triangle but with the inequality reversed for an obtuse triangle.

The median m_c from the longest side is greater or less than the circumradius for an acute or obtuse triangle respectively:

$$m_c > R$$

for acute triangles, with the opposite for obtuse triangles.

Area

Ono's inequality for the area A,

$$27(b^2 + c^2 - a^2)^2 (c^2 + a^2 - b^2)^2 (a^2 + b^2 - c^2)^2 \le (4A)^6,$$

holds for all acute triangles but not for all obtuse triangles.

Trigonometric Functions

For an acute triangle we have, for angles A, B, and C,

$$\cos^2 A + \cos^2 B + \cos^2 C < 1,$$

with the reverse inequality holding for an obtuse triangle.

For an acute triangle with circumradius R,

$$a\cos^3 A + b\cos^3 B + c\cos^3 C \le \frac{abc}{4R^2}$$

and,

$$\cos^3 A + \cos^3 B + \cos^3 C + \cos A \cos B \cos C \ge \frac{1}{2}.$$

For an acute triangle,

$$\sin^2 A + \sin^2 B + \sin^2 C > 2,$$

with the reverse inequality for an obtuse triangle.

For an acute triangle,

$$\sin A \cdot \sin B + \sin B \cdot \sin C + \sin C \cdot \sin A \le (\cos A + \cos B + \cos C)^2.$$

For any triangle the triple tangent identity states that the sum of the angles' tangents equals their product. Since an acute angle has a positive tangent value while an obtuse angle has a negative one, the expression for the product of the tangents shows that:

$$\tan A + \tan B + \tan C = \tan A \cdot \tan B \cdot \tan C > 0$$

for acute triangles, while the opposite direction of inequality holds for obtuse triangles.

We have,

$$\tan A + \tan B + \tan C \ge 2(\sin 2A + \sin 2B + \sin 2C)$$

for acute triangles, and the reverse for obtuse triangles.

For all acute triangles,

$$(\tan A + \tan B + \tan C)^2 \ge (\sec A + 1)^2 + (\sec B + 1)^2 + (\sec C + 1)^2.$$

For all acute triangles with inradius r and circumradius R,

$$a \tan A + b \tan B + c \tan C \ge 10R - 2r.$$

For an acute triangle with area K,

$$(\sqrt{\cot A} + \sqrt{\cot B} + \sqrt{\cot C})^2 \le \frac{K}{r^2}.$$

Circumradius, Inradius and Exradii

In an acute triangle, the sum of the circumradius R and the inradius r is less than half the sum of the shortest sides a and b:

$$R + r < \frac{a+b}{2},$$

while the reverse inequality holds for an obtuse triangle.

For an acute triangle with medians m_a, m_b, and m_c and circumradius R, we have,

$$m_a^2 + m_b^2 + m_c^2 > 6R^2$$

while the opposite inequality holds for an obtuse triangle.

Also, an acute triangle satisfies:

$$r^2 + r_a^2 + r_b^2 + r_c^2 < 8R^2,$$

in terms of the excircle radii r_a, r_b, and r_c, again with the reverse inequality holding for an obtuse triangle.

For an acute triangle with semiperimeter s,

$$s - r > 2R,$$

and the reverse inequality holds for an obtuse triangle.

For an acute triangle with area K,

$$ab + bc + ca \geq 2R(R + r) + \frac{8K}{\sqrt{3}}.$$

Distances involving Triangle Centers

For an acute triangle the distance between the circumcenter O and the orthocenter H satisfies:

$$OH < R,$$

with the opposite inequality holding for an obtuse triangle.

For an acute triangle the distance between the incircle center I and orthocenter H satisfies:

$$IH < r\sqrt{2},$$

where r is the inradius, with the reverse inequality for an obtuse triangle.

Inscribed Square

If one of the inscribed squares of an acute triangle has side length x_a and another has side length x_b with $x_a < x_b$, then,

$$1 \geq \frac{x_a}{x_b} \geq \frac{2\sqrt{2}}{3} \approx 0.94.$$

Two Triangles

If two obtuse triangles have sides (a, b, c) and (p, q, r) with c and r being the respective longest sides, then,

$$ap + bq < cr.$$

Triangles with Special Names

The Calabi triangle, which is the only non-equilateral triangle for which the largest square that fits in the interior can be positioned in any of three different ways, is obtuse and isosceles with base angles 39.1320261...° and third angle 101.7359477°.

- The equilateral triangle, with three 60° angles, is acute.

- The Morley triangle, formed from any triangle by the intersections of its adjacent angle trisectors, is equilateral and hence acute.

- The golden triangle is the isosceles triangle in which the ratio of the duplicated side to the base side equals the golden ratio. It is acute, with angles 36°, 72°, and 72°, making it the only triangle with angles in the proportions 1:2:2.

- The heptagonal triangle, with sides coinciding with a side, the shorter diagonal, and the longer diagonal of a regular heptagon, is obtuse, with angles $\pi/7, 2\pi/7, and\, 4\pi/7$.

Triangles with Integer Sides

The only triangle with consecutive integers for an altitude and the sides is acute, having sides (13,14,15) and altitude from side 14 equal to 12.

The smallest-perimeter triangle with integer sides in arithmetic progression, and the smallest-perimeter integer-sided triangle with distinct sides, is obtuse: namely the one with sides (2, 3, 4).

The only triangles with one angle being twice another and having integer sides in arithmetic progression are acute: namely, the (4,5,6) triangle and its multiples.

There are no acute integer-sided triangles with area = perimeter, but there are three obtuse ones, having sides (6,25,29), (7,15,20), and (9,10,17).

The smallest integer-sided triangle with three rational medians is acute, with sides (68, 85, 87).

Heron triangles have integer sides and integer area. The oblique Heron triangle with the smallest perimeter is acute, with sides (6, 5, 5). The two oblique Heron triangles that share the smallest area are the acute one with sides (6, 5, 5) and the obtuse one with sides (8, 5, 5), the area of each being 12.

RIGHT TRIANGLE

A right triangle or right-angled triangle is a triangle in which one angle is a right angle (that is, a 90-degree angle). The relation between the sides and angles of a right triangle is the basis for trigonometry.

The side opposite the right angle is called the hypotenuse (side c in the figure). The sides adjacent to the right angle are called legs (or catheti). Side a may be identified as the side adjacent to angle B and opposed to (or opposite) angle A, while side b is the side adjacent to angle A and opposed to angle B.

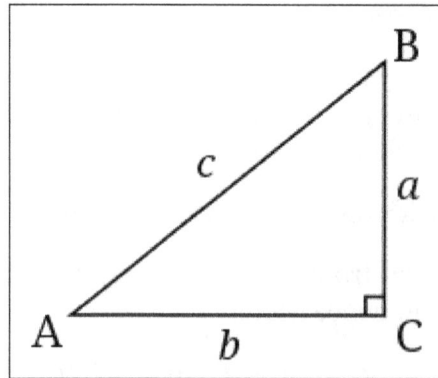

If the lengths of all three sides of a right triangle are integers, the triangle is said to be a Pythagorean triangle and its side lengths are collectively known as a Pythagorean triple.

Principal Properties

Area

As with any triangle, the area is equal to one half the base multiplied by the corresponding height. In a right triangle, if one leg is taken as the base then the other is height, so the area of a right triangle is one half the product of the two legs. As a formula the area T is:

$$T = \tfrac{1}{2} ab$$

where a and b are the legs of the triangle.

If the incircle is tangent to the hypotenuse AB at point P, then denoting the semi-perimeter $(a + b + c) / 2$ as s, we have PA = $s - a$ and PB = $s - b$, and the area is given by:

$$T = \mathrm{PA} \cdot \mathrm{PB} = (s-a)(s-b).$$

This formula only applies to right triangles.

Altitudes

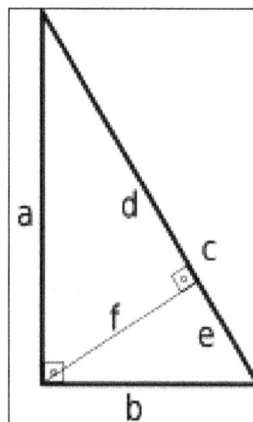

Altitude of a right triangle.

If an altitude is drawn from the vertex with the right angle to the hypotenuse then the triangle is divided into two smaller triangles which are both similar to the original and therefore similar to each other. From this:

- The altitude to the hypotenuse is the geometric mean (mean proportional) of the two segments of the hypotenuse.

- Each leg of the triangle is the mean proportional of the hypotenuse and the segment of the hypotenuse that is adjacent to the leg.

In equations,

$$f^2 = de, \text{ (this is sometimes known as the right triangle altitude theorem)}$$

$$b^2 = ce,$$

$$a^2 = cd$$

where a, b, c, d, e, f are as shown in the diagram. Thus,

$$f = \frac{ab}{c}.$$

Moreover, the altitude to the hypotenuse is related to the legs of the right triangle by:

$$\frac{1}{a^2} + \frac{1}{b^2} = \frac{1}{f^2}.$$

For solutions of this equation in integer values of $a, b, f,$ and c.

The altitude from either leg coincides with the other leg. Since these intersect at the right-angled vertex, the right triangle's orthocenter—the intersection of its three altitudes—coincides with the right-angled vertex.

Pythagorean Theorem

The Pythagorean theorem states that:

In any right triangle, the area of the square whose side is the hypotenuse (the side opposite the right angle) is equal to the sum of the areas of the squares whose sides are the two legs (the two sides that meet at a right angle).

This can be stated in equation form as:

$$a^2 + b^2 = c^2$$

where c is the length of the hypotenuse, and a and b are the lengths of the remaining two sides.

Pythagorean triples are integer values of a, b, c satisfying this equation.

Inradius and Circumradius

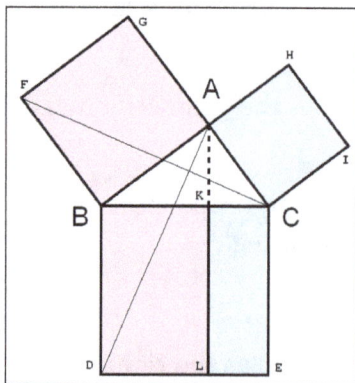

Illustration of the Pythagorean Theorem.

The radius of the incircle of a right triangle with legs a and b and hypotenuse c is:

$$r = \frac{a+b-c}{2} = \frac{ab}{a+b+c}.$$

The radius of the circumcircle is half the length of the hypotenuse,

$$R = \frac{c}{2}.$$

Thus the sum of the circumradius and the inradius is half the sum of the legs:

$$R + r = \frac{a\ \ b}{}$$

One of the legs can be expressed in terms of the inradius and the other leg as:

$$a = \frac{2r(b-r)}{b-2r}.$$

Characterizations

A triangle ABC with sides $a \leq b < c$, semiperimeter s, area T, altitude h opposite the longest side, circumradius R, inradius r, exradii r_a, r_b, r_c (tangent to a, b, c respectively), and medians m_a, m_b, m_c is a right triangle if and only if any one of the statements in the following six categories is true. All of them are of course also properties of a right triangle, since characterizations are equivalences.

Sides and Semiperimeter

- $a^2 + b^2 = c^2$ (Pythagorean theorem).

- $(s-a)(s-b) = s(s-c)$.

- $s = 2R + r.s = 2R + r.$

- $a^2 + b^2 + c^2 = 8R^2.$

Angles

- A and B are complementary.

- $\cos A \cos B \cos C = 0.$

- $\sin^2 A + \sin^2 B + \sin^2 C = 2.$

- $\cos^2 A + \cos^2 B + \cos^2 C = 1.$

- $\sin 2A = \sin 2B = 2 \sin A \sin B.$

Area

- $T = \dfrac{ab}{2}$

- $T = r_a r_b = r r_c$

- $T = r(2R + r)$

- $T = PA \cdot PB$, where P is the tangency point of the incircle at the longest side AB.

Inradius and Exradii

- $r = s - c = (a + b - c)/2$

- $r_a = s - b = (a - b + c)/2$

- $r_b = s - a = (-a + b + c)/2$

- $r_c = s = (a + b + c)/2$

- $r_a + r_b + r_c + r = a + b + c$

- $r_a^2 + r_b^2 + r_c^2 + r^2 = a^2 + b^2 + c^2$

- $r = \dfrac{r_a r_b}{r_c}.$

Altitude and Medians

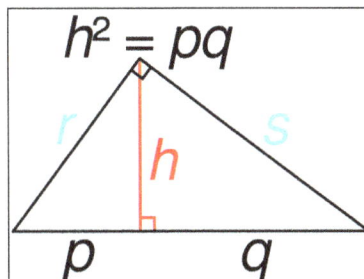

The altitude of a right triangle from its right angle to its hypotenuse is the geometric mean of the lengths of the segments the hypotenuse is split into. Using Pythagoras' theorem on the 3 triangles of sides $(p+q,r,s)$, (r,h,p) and (s,h,q), $(p+q)^2 = r^2 + s^2$ $p^2 + 2pq + q^2 = (h^2+p^2)+(h^2+q^2)$ $2pq = 2h^2$.

- $h = \dfrac{ab}{c}$

- $m_a^2 + m_b^2 + m_c^2 = 6R^2$.

- The length of one median is equal to the circumradius.

- The shortest altitude (the one from the vertex with the biggest angle) is the geometric mean of the line segments it divides the opposite (longest) side into. This is the right triangle altitude theorem.

Circumcircle and Incircle

- The triangle can be inscribed in a semicircle, with one side coinciding with the entirety of the diameter (Thales' theorem).

- The circumcenter is the midpoint of the longest side.

- The longest side is a diameter of the circumcircle ($c = 2R$).

- The circumcircle is tangent to the nine-point circle.

- The orthocenter lies on the circumcircle.

- The distance between the incenter and the orthocenter is equal to $\sqrt{2}r$.

Trigonometric Ratios

The trigonometric functions for acute angles can be defined as ratios of the sides of a right triangle. For a given angle, a right triangle may be constructed with this angle, and the sides labeled opposite, adjacent and hypotenuse with reference to this angle according to the definitions above. These ratios of the sides do not depend on the particular right triangle chosen, but only on the given angle, since all triangles constructed this way are similar. If, for a given angle α, the opposite side, adjacent side and hypotenuse are labeled O, A and H respectively, then the trigonometric functions are:

$$\sin \alpha = \frac{O}{H}, \cos \alpha = \frac{A}{H}, \tan \alpha = \frac{O}{A}, \sec \alpha = \frac{H}{A}, \cot \alpha = \frac{A}{O}, \csc \alpha = \frac{H}{O}.$$

For the expression of hyperbolic functions as ratio of the sides of a right triangle.

Special Right Triangles

The values of the trigonometric functions can be evaluated exactly for certain angles using right triangles with special angles. These include the *30-60-90 triangle* which can be used to evaluate the trigonometric functions for any multiple of $\pi/6$, and the *45-45-90 triangle* which can be used to evaluate the trigonometric functions for any multiple of $\pi/4$.

Kepler Triangle

Let H, G, and A be the harmonic mean, the geometric mean, and the arithmetic mean of two positive numbers a and b with $a > b$. If a right triangle has legs H and G and hypotenuse A, then:

$$\frac{A}{H} = \frac{A^2}{G^2} = \frac{G^2}{H^2} = \phi$$

and,

$$\frac{a}{b} = \phi^3,$$

where ϕ is the golden ratio $\frac{1+\sqrt{5}}{2}$. Since the sides of this right triangle are in geometric progression, this is the Kepler triangle.

Thales' Theorem

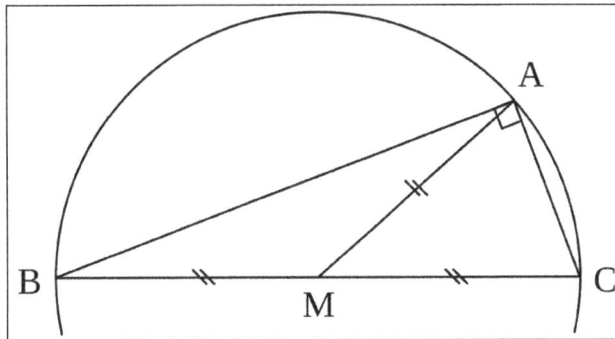

Median of a right angle of a triangle.

Thales' theorem states that if A is any point of the circle with diameter BC (except B or C themselves) ABC is a right triangle where A is the right angle. The converse states that if a right triangle is inscribed in a circle then the hypotenuse will be a diameter of the circle. A corollary is that the length of the hypotenuse is twice the distance from the right angle vertex to the midpoint of the hypotenuse. Also, the center of the circle that circumscribes a right triangle is the midpoint of the hypotenuse and its radius is one half the length of the hypotenuse.

Medians

The following formulas hold for the medians of a right triangle:

$$m_a^2 + m_b^2 = 5m_c^2 = \frac{5}{4}c^2.$$

The median on the hypotenuse of a right triangle divides the triangle into two isosceles triangles, because the median equals one-half the hypotenuse.

The medians m_a and m_b from the legs satisfy:

$$4c^4 + 9a^2b^2 = 16m_a^2 m_b^2.$$

Euler Line

In a right triangle, the Euler line contains the median on the hypotenuse—that is, it goes through both the right-angled vertex and the midpoint of the side opposite that vertex. This is because the right triangle's orthocenter, the intersection of its altitudes, falls on the right-angled vertex while its circumcenter, the intersection of its perpendicular bisectors of sides, falls on the midpoint of the hypotenuse.

Inequalities

In any right triangle the diameter of the incircle is less than half the hypotenuse, and more strongly it is less than or equal to the hypotenuse times $(\sqrt{2} - 1)$.

In a right triangle with legs a, b and hypotenuse c,

$$c \geq \frac{\sqrt{2}}{2}(a+b)$$

with equality only in the isosceles case.

If the altitude from the hypotenuse is denoted h_c, then:

$$h_c \leq \frac{\sqrt{2}}{4}(a+b)$$

with equality only in the isosceles case.

Other Properties

If segments of lengths p and q emanating from vertex C trisect the hypotenuse into segments of length $c/3$, then:

$$p^2 + q^2 = 5\left(\frac{c}{3}\right)^2.$$

The right triangle is the only triangle having two, rather than one or three, distinct inscribed squares.

Let h and k ($h > k$) be the sides of the two inscribed squares in a right triangle with hypotenuse c. Then:

$$\frac{1}{c^2} + \frac{1}{h^2} = \frac{1}{k^2}.$$

These sides and the incircle radius r are related by a similar formula:

$$\frac{1}{r} = -\frac{1}{c} + \frac{1}{h} + \frac{1}{k}.$$

The perimeter of a right triangle equals the sum of the radii of the incircle and the three excircles:

$$a + b + c = r + r_a + r_b + r_c.$$

RECTANGLE

A Rectangle is a four sided-quadrilateral having all the internal angles to be right-angled (90o).

It is to be noted that in a rectangle the opposite sides are equal in length which makes it different from a square.

For example, if one side of a rectangle is 20 cm, then the side opposite to it is also 20 cm.

Rectangle

A rectangle is characterized by length (L) and width (W). Both length and width are different in size.

In the figure above, a rectangle ABCD has four sides as AB, BC, CD, and DA and right angles A, B, C, and D. The distance between A and B or C and D is defined as the length (L), whereas the distance between B and C or A and D is defined as Width (W) of the given rectangle.

Real World Application of Rectangles

Table, Book, TV screen, Mobile phone, Wall, Magazine, Tennis court, etc.

Perimeter of a Rectangle

The perimeter is defined as the total distance around the surface.

Mathematically Perimeter is given as:

P=2(*Length+Width*) unit length

Area of Rectangle

Before calculating the area of a rectangle, let us know what exactly an area is. An area is a way of measuring how much space is contained inside a particular figure.

Mathematically, it is given as:

$A = Length \times Width \; unit^2$

List of Properties of Rectangle

To understand the properties of a rectangle, let's learn what is a rectangle first. As we have seen many different shapes which are either 2 dimensional or 3 dimensional and so on. Out of these shapes, Rectangle is an important two-dimensional shape. It is a four-sided polygon. All the four sides are straight with four right angles i.e each angle is a right angle in a rectangle. The most common everyday things we see and is rectangular in shape is Television, computer screen, notebook, mobile phones, CPU, Notice boards etc.

- It's a parallelogram with four right angles.

- It's diagonals bisect each other.

- The opposite side of a rectangle is equal.

- The opposite side of the rectangle is parallel.

- The area of a recangle is = l x b, where l = length and b = breadth.

- The perimeter of a rectangle is 2(l + b).

- The sum of all the interior angle of a rectangle is 3600.

- The diagonal of a recatngle acts like a diameter of its circumference.

Diagonal of a Rectangle

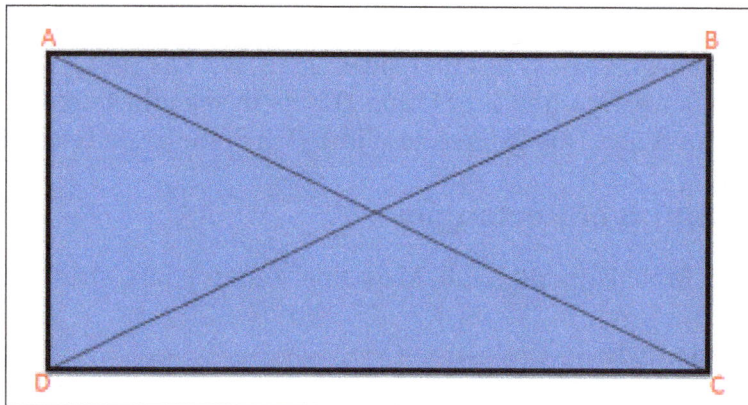

Rectangle- Diagonals.

A rectangle has two diagonals, they are equal in length and intersect in the middle.

Length of Diagonals

The length of diagonals can be found using the Pythagoras Theorem:

$$D = \sqrt{L^2 + W^2}$$

SQUARE

In geometry, a square is a regular quadrilateral, which means that it has four equal sides and four equal angles (90-degree angles, or (100-gradian angles or right angles). It can also be defined as a rectangle in which two adjacent sides have equal length. A square with vertices *ABCD* would be denoted □*ABCD*.

Characterizations

A convex quadrilateral is a square if and only if it is any one of the following:

- A rectangle with two adjacent equal sides.

- A rhombus with a right vertex angle.

- A rhombus with all angles equal.

- A parallelogram with one right vertex angle and two adjacent equal sides.

- A quadrilateral with four equal sides and four right angles.

- A quadrilateral where the diagonals are equal and are the perpendicular bisectors of each other, i.e. a rhombus with equal diagonals.

- A convex quadrilateral with successive sides a, b, c, d whose area is $A = \frac{1}{2}(a^2 + c^2) = \frac{1}{2}(b^2 + d^2)$.

Properties

A square is a special case of a rhombus (equal sides, opposite equal angles), a kite (two pairs of adjacent equal sides), a trapezoid (one pair of opposite sides parallel), a parallelogram (all opposite sides parallel), a quadrilateral or tetragon (four-sided polygon), and a rectangle (opposite sides equal, right-angles) and therefore has all the properties of all these shapes, namely:

- The diagonals of a square bisect each other and meet at 90°.

- The diagonals of a square bisect its angles.

- Opposite sides of a square are both parallel and equal in length.

- All four angles of a square are equal. (Each is 360°/4 = 90°, so every angle of a square is a right angle).

- All four sides of a square are equal.

- The diagonals of a square are equal.

- The square is the n=2 case of the families of n-hypercubes and n-orthoplexes.

- A square has Schläfli symbol {4}. A truncated square, t{4}, is an octagon, {8}. An alternated square, h{4}, is a digon, {2}.

Perimeter and Area

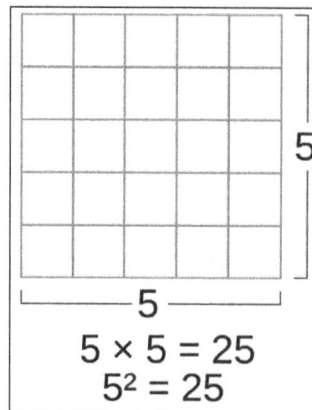

The area of a square is the product of the length of its sides.

The perimeter of a square whose four sides have length ℓ is:

$$P = 4\ell$$

and the area A is:

$$A = \ell^2.$$

In classical times, the second power was described in terms of the area of a square, as in the above formula. This led to the use of the term *square* to mean raising to the second power.

The area can also be calculated using the diagonal d according to:

$$A = \frac{d^2}{2}.$$

In terms of the circumradius R, the area of a square is:

$$A = 2R^2;$$

since the area of the circle is πR^2, the square fills approximately 0.6366 of its circumscribed circle.

In terms of the inradius r, the area of the square is:

$$A = 4r^2.$$

Because it is a regular polygon, a square is the quadrilateral of least perimeter enclosing a given area. Dually, a square is the quadrilateral containing the largest area within a given perimeter. Indeed, if A and P are the area and perimeter enclosed by a quadrilateral, then the following isoperimetric inequality holds:

$$16A \leq P^2$$

with equality if and only if the quadrilateral is a square.

- The diagonals of a square are $\sqrt{2}$ (about 1.414) times the length of a side of the square. This value, known as the square root of 2 or Pythagoras' constant, was the first number proven to be irrational.

- A square can also be defined as a parallelogram with equal diagonals that bisect the angles.

- If a figure is both a rectangle (right angles) and a rhombus (equal edge lengths), then it is a square.

- If a circle is circumscribed around a square, the area of the circle is $\pi/2$ (about 1.5708) times the area of the square.

- If a circle is inscribed in the square, the area of the circle is $\pi/4$ (about 0.7854) times the area of the square.

- A square has a larger area than any other quadrilateral with the same perimeter.

- A square tiling is one of three regular tilings of the plane (the others are the equilateral triangle and the regular hexagon).

- The square is in two families of polytopes in two dimensions: hypercube and the cross-polytope. The Schläfli symbol for the square is {4}.

- The square is a highly symmetric object. There are four lines of reflectional symmetry and it has rotational symmetry of order 4 (through 90°, 180° and 270°). Its symmetry group is the dihedral group D_4.

- If the inscribed circle of a square $ABCD$ has tangency points E on AB, F on BC, G on CD, and H on DA, then for any point P on the inscribed circle:

$$2(PH^2 - PE^2) = PD^2 - PB^2.$$

- If d_i is the distance from an arbitrary point in the plane to the i-th vertex of a square and R is the circumradius of the square, then:

$$\frac{d_1^4 + d_2^4 + d_3^4 + d_4^4}{4} + 3R^4 = \left(\frac{d_1^2 + d_2^2 + d_3^2 + d_4^2}{4} + R^2\right)^2.$$

Coordinates and Equations

The coordinates for the vertices of a square with vertical and horizontal sides, centered at the origin and with side length 2 are (± 1, ± 1), while the interior of this square consists of all points (x_i, y_i) with $-1 < x_i < 1$ and $-1 < y_i < 1$. The equation:

$$\max(x^2, y^2) = 1$$

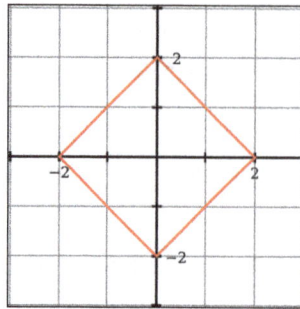

$|x|+|y|=2$ plotted on *Cartesian coordinates*.

specifies the boundary of this square. This equation means "x^2 or y^2, whichever is larger, equals 1." The circumradius of this square (the radius of a circle drawn through the square's vertices) is half the square's diagonal, and equals $\sqrt{2}$.

Then the circumcircle has the equation:

$$x^2 + y^2 = 2.$$

Alternatively the equation:

$$|x-a| + |y-b| = r.$$

can also be used to describe the boundary of a square with center coordinates (a, b) and a horizontal or vertical radius of r.

Symmetry

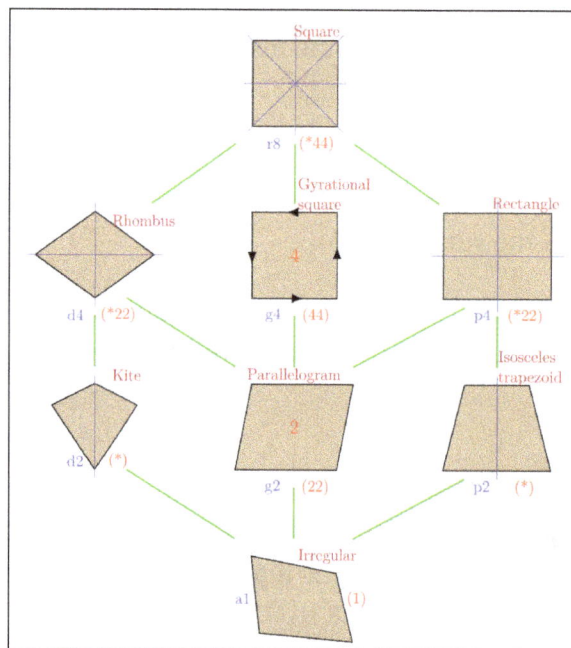

The dihedral symmetries are divided depending on whether they pass through vertices (d for diagonal) or edges (p for perpendiculars) Cyclic symmetries in the middle column are labeled as g for their central gyration orders. Full symmetry of the square is r12 and no symmetry is labeled a1.

The *square* has Dih$_4$ symmetry, order 8. There are 2 dihedral subgroups: Dih$_2$, Dih$_1$, and 3 cyclic subgroups: Z$_4$, Z$_2$, and Z$_1$.

A square is a special case of many lower symmetry quadrilaterals:

- A rectangle with two adjacent equal sides.

- A quadrilateral with four equal sides and four right angles.

- A parallelogram with one right angle and two adjacent equal sides.

- A rhombus with a right angle.

- A rhombus with all angles equal.

- A rhombus with equal diagonals.

These 6 symmetries express 8 distinct symmetries on a square. John Conway labels these by a letter and group order.

Each subgroup symmetry allows one or more degrees of freedom for irregular quadrilaterals. r8 is full symmetry of the square, and a1 is no symmetry. d4, is the symmetry of a rectangle and p4, is the symmetry of a rhombus. These two forms are duals of each other and have half the symmetry order of the square. d2 is the symmetry of an isosceles trapezoid, and p2 is the symmetry of a kite. g2 defines the geometry of a parallelogram.

Only the g4 subgroup has no degrees of freedom but can seen as a square with directed edges.

Squares Inscribed in Triangles

Every acute triangle has three inscribed squares (squares in its interior such that all four of a square's vertices lie on a side of the triangle, so two of them lie on the same side and hence one side of the square coincides with part of a side of the triangle). In a right triangle two of the squares coincide and have a vertex at the triangle's right angle, so a right triangle has only two *distinct* inscribed squares. An obtuse triangle has only one inscribed square, with a side coinciding with part of the triangle's longest side.

The fraction of the triangle's area that is filled by the square is no more than 1/2.

Squaring the Circle

Squaring the circle is the problem, proposed by ancient geometers, of constructing a square with the same area as a given circle by using only a finite number of steps with compass and straightedge.

In 1882, the task was proven to be impossible, as a consequence of the Lindemann–Weierstrass theorem which proves that pi (π) is a transcendental number, rather than an algebraic irrational number; that is, it is not the root of any polynomial with rational coefficients.

Non-Euclidean Geometry

In non-Euclidean geometry, squares are more generally polygons with 4 equal sides and equal angles.

In spherical geometry, a square is a polygon whose edges are great circle arcs of equal distance, which meet at equal angles. Unlike the square of plane geometry, the angles of such a square are larger than a right angle. Larger spherical squares have larger angles.

In hyperbolic geometry, squares with right angles do not exist. Rather, squares in hyperbolic geometry have angles of less than right angles. Larger hyperbolic squares have smaller angles.

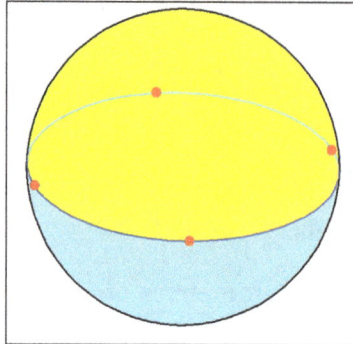

Two squares can tile the sphere with 2 squares around each vertex and 180-degree internal angles. Each square covers an entire hemisphere and their vertices lie along a great circle. This is called a spherical square dihedron. The Schläfli symbol is {4,2}.

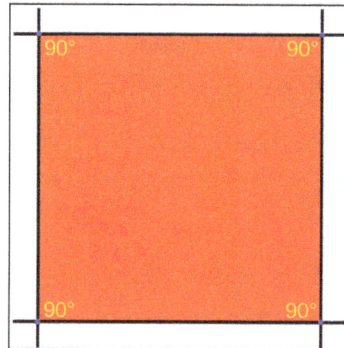

Squares can tile the Euclidean plane with 4 around each vertex, with each square having an internal angle of 90°. The Schläfli symbol is {4,4}.

Crossed Square

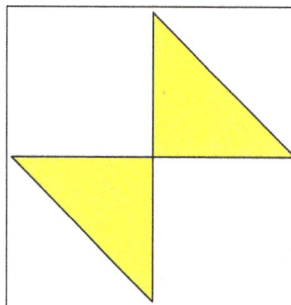

Crossed-square.

A crossed square is a faceting of the square, a self-intersecting polygon created by removing two opposite edges of a square and reconnecting by its two diagonals. It has half the symmetry of the square, Dih_2, order 4. It has the same vertex arrangement as the square, and is vertex-transitive.

It appears as two 45-45-90 triangle with a common vertex, but the geometric intersection is not considered a vertex.

A crossed square is sometimes likened to a bow tie or butterfly. the crossed rectangle is related, as a faceting of the rectangle, both special cases of crossed quadrilaterals.

The interior of a crossed square can have a polygon density of ±1 in each triangle, dependent upon the winding orientation as clockwise or counterclockwise.

A square and a crossed square have the following properties in common:

- Opposite sides are equal in length.

- The two diagonals are equal in length.

- It has two lines of reflectional symmetry and rotational symmetry of order 2 (through 180°).

It exists in the vertex figure of a uniform star polyhedra, the tetrahemihexahedron.

Graphs

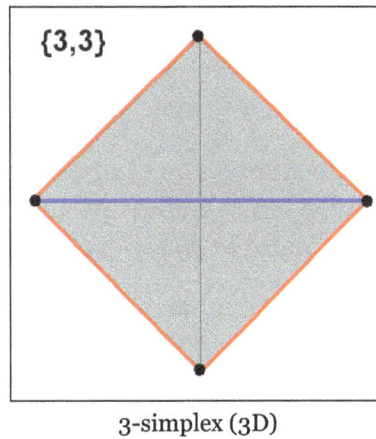

3-simplex (3D)

The K_4 complete graph is often drawn as a square with all 6 possible edges connected, hence appearing as a square with both diagonals drawn. This graph also represents an orthographic projection of the 4 vertices and 6 edges of the regular 3-simplex (tetrahedron).

RHOMBUS

In plane Euclidean geometry, a rhombus (plural rhombi or rhombuses) is a simple (non-self-intersecting) quadrilateral whose four sides all have the same length. Another name is equilateral quadrilateral, since equilateral means that all of its sides are equal in length. The rhombus is often called a diamond, after the diamonds suit in playing cards which resembles the projection of an octahedral diamond, or a lozenge, though the former sometimes refers specifically to a rhombus with a 60° angle, and the latter sometimes refers specifically to a rhombus with a 45° angle.

Every rhombus is a parallelogram and a kite. A rhombus with right angles is a square.

Characterizations

A simple (non-self-intersecting) quadrilateral is a rhombus if and only if it is any one of the following:

- A parallelogram in which a diagonal bisects an interior angle.

- A parallelogram in which at least two consecutive sides are equal in length.

- A parallelogram in which the diagonals are perpendicular (an orthodiagonal parallelogram).

- A quadrilateral with four sides of equal length (by definition).

- A quadrilateral in which the diagonals are perpendicular and bisect each other.

- A quadrilateral in which each diagonal bisects two opposite interior angles.

- A quadrilateral *ABCD* possessing a point *P* in its plane such that the four triangles *ABP*, *BCP*, *CDP*, and *DAP* are all congruent.

- A quadrilateral *ABCD* in which the incircles in triangles *ABC*, *BCD*, *CDA* and *DAB* have a common point.

Basic Properties

Every rhombus has two diagonals connecting pairs of opposite vertices, and two pairs of parallel sides. Using congruent triangles, one can prove that the rhombus is symmetric across each of these diagonals. It follows that any rhombus has the following properties:

- Opposite angles of a rhombus have equal measure.

- The two diagonals of a rhombus are perpendicular; that is, a rhombus is an orthodiagonal quadrilateral.

- Its diagonals bisect opposite angles.

The first property implies that every rhombus is a parallelogram. A rhombus therefore has all of the properties of a parallelogram: for example, opposite sides are parallel; adjacent angles are supplementary; the two diagonals bisect one another; any line through the midpoint bisects the area; and the sum of the squares of the sides equals the sum of the squares of the diagonals (the parallelogram law). Thus denoting the common side as *a* and the diagonals as *p* and *q*, in every rhombus:

$$4a^2 = p^2 + q^2.$$

Not every parallelogram is a rhombus, though any parallelogram with perpendicular diagonals (the second property) is a rhombus. In general, any quadrilateral with perpendicular diagonals, one of which is a line of symmetry, is a kite. Every rhombus is a kite, and any quadrilateral that is both a kite and parallelogram is a rhombus.

A rhombus is a tangential quadrilateral. That is, it has an inscribed circle that is tangent to all four sides.

Diagonals

The length of the diagonals $p = AC$ and $q = BD$ can be expressed in terms of the rhombus side a and one vertex angle α as:

$$p = a\sqrt{2 + 2\cos\alpha}$$

and,

$$q = a\sqrt{2 - 2\cos\alpha}.$$

These formulas are a direct consequence of the law of cosines.

Inradius

The inradius (the radius of a circle inscribed in the rhombus), denoted by r, can be expressed in terms of the diagonals p and q as:

$$r = \frac{p \cdot q}{2\sqrt{p^2 + q^2}}.$$

or in terms of the side length a and any vertex angle α or β as:

$$r = \frac{a\sin\alpha}{2} = \frac{a\sin\beta}{2}.$$

Area

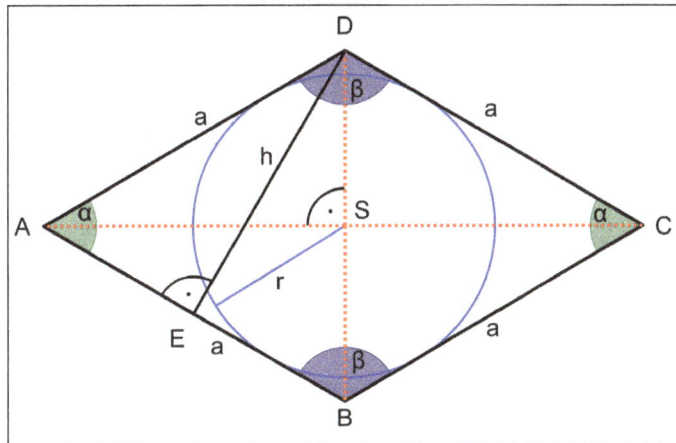

A rhombus. Each angle marked with a black dot is a right angle. The height h is the perpendicular distance between any two non-adjacent sides, which equals the diameter of the circle inscribed. The diagonals of lengths p and q are the red dotted line segments.

As for all parallelograms, the area K of a rhombus is the product of its base and its height (h). The base is simply any side length a:

$$K = a \cdot h.$$

The area can also be expressed as the base squared times the sine of any angle:

$$K = a^2 \cdot \sin\alpha = a^2 \cdot \sin\beta$$

or in terms of the height and a vertex angle:

$$K = \frac{h^2}{\sin\alpha},$$

or as half the product of the diagonals p, q:

$$K = \frac{p \cdot q}{2},$$

or as the semiperimeter times the radius of the circle inscribed in the rhombus (inradius):

$$K = 2a \cdot r.$$

Another way, in common with parallelograms, is to consider two adjacent sides as vectors, forming a bivector, so the area is the magnitude of the bivector (the magnitude of the vector product of the two vectors), which is the determinant of the two vectors' Cartesian coordinates: $K = x_1 y_2 - x_2 y_1$.

Dual Properties

The dual polygon of a rhombus is a rectangle:

- A rhombus has all sides equal, while a rectangle has all angles equal.

- A rhombus has opposite angles equal, while a rectangle has opposite sides equal.

- A rhombus has an inscribed circle, while a rectangle has a circumcircle.

- A rhombus has an axis of symmetry through each pair of opposite vertex angles, while a rectangle has an axis of symmetry through each pair of opposite sides.

- The diagonals of a rhombus intersect at equal angles, while the diagonals of a rectangle are equal in length.

- The figure formed by joining the midpoints of the sides of a rhombus is a rectangle and vice versa.

Cartesian Equation

The sides of a rhombus centered at the origin, with diagonals each falling on an axis, consist of all points (x, y) satisfying:

$$\left|\frac{x}{a}\right| + \left|\frac{y}{b}\right| = 1.$$

The vertices are at $(\pm a, 0)$ and $(0, \pm b)$. This is a special case of the superellipse, with exponent 1.

Other Properties

- One of the five 2D lattice types is the rhombic lattice, also called centered rectangular lattice.

- Identical rhombi can tile the 2D plane in three different ways, including, for the 60° rhombus, the rhombille tiling.

- Three-dimensional analogues of a rhombus include the bipyramid and the bicone.

- Several polyhedra have rhombic faces, such as the rhombic dodecahedron and the trapezo-rhombic dodecahedron.

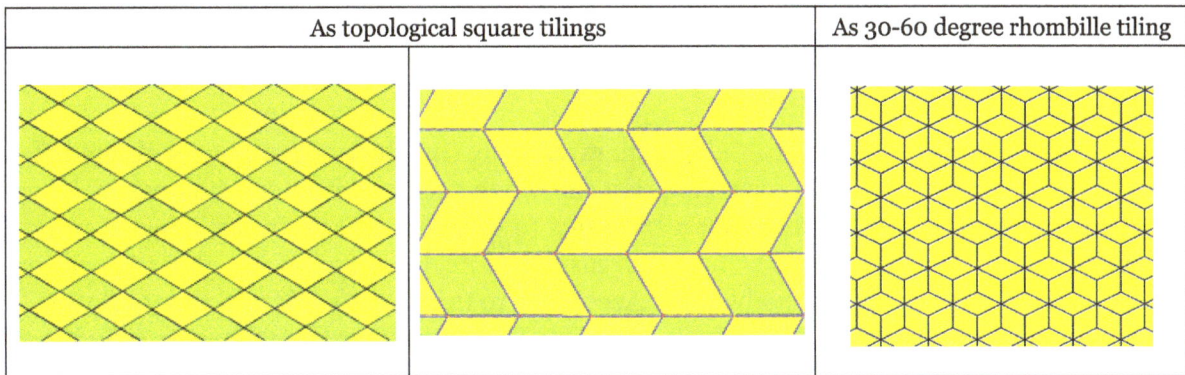

As topological square tilings		As 30-60 degree rhombille tiling

As the Faces of a Polyhedron

- A rhombohedron is a three-dimensional figure like a cube, except that its six faces are rhombi instead of squares.

- The rhombic dodecahedron is a convex polyhedron with 12 congruent rhombi as its faces.

- The rhombic triacontahedron is a convex polyhedron with 30 golden rhombi (rhombi whose diagonals are in the golden ratio) as its faces.

- The great rhombic triacontahedron is a nonconvex isohedral, isotoxal polyhedron with 30 intersecting rhombic faces.

- The rhombic hexecontahedron is a stellation of the rhombic triacontahedron. It is nonconvex with 60 golden rhombic faces with icosahedral symmetry.

- The rhombic enneacontahedron is a polyhedron composed of 90 rhombic faces, with three, five, or six rhombi meeting at each vertex. It has 60 broad rhombi and 30 slim ones.

- The trapezo-rhombic dodecahedron is a convex polyhedron with 6 rhombic and 6 trapezoidal faces.

- The rhombic icosahedron is a polyhedron composed of 20 rhombic faces, of which three, four, or five meet at each vertex. It has 10 faces on the polar axis with 10 faces following the equator.

KITE

In Euclidean geometry, a kite is a quadrilateral whose four sides can be grouped into two pairs of equal-length sides that are adjacent to each other. In contrast, a parallelogram also has two pairs of equal-length sides, but they are opposite to each other rather than adjacent. Kite quadrilaterals are named for the wind-blown, flying kites, which often have this shape and which are in turn named for a bird. Kites are also known as deltoids, but the word "deltoid" may also refer to a deltoid curve, an unrelated geometric object.

A kite, as defined above, may be either convex or concave, but the word "kite" is often restricted to the convex variety. A concave kite is sometimes called a "dart" or "arrowhead", and is a type of pseudotriangle.

It is possible to classify quadrilaterals either hierarchically (in which some classes of quadrilaterals are subsets of other classes) or as a partition (in which each quadrilateral belongs to only one class). With a hierarchical classification, a rhombus (a quadrilateral with four sides of the same length) or a square is considered to be a special case of a kite, because it is possible to partition its edges into two adjacent pairs of equal length. According to this classification, every equilateral kite is a rhombus, and every equiangular kite is a square. However, with a partitioning classification, rhombi and squares are not considered to be kites, and it is not possible for a kite to be equilateral or equiangular. For the same reason, with a partitioning classification, shapes meeting the additional constraints of other classes of quadrilaterals, such as the right kites discussed below, would not be considered to be kites.

A kite with three equal 108° angles and one 36° angle forms the convex hull of the lute of Pythagoras.

The kites that are also cyclic quadrilaterals (i.e. the kites that can be inscribed in a circle) are exactly the ones formed from two congruent right triangles. That is, for these kites the two equal angles on opposite sides of the symmetry axis are each 90 degrees. These shapes are called right kites. Because they circumscribe one circle and are inscribed in another circle, they are bicentric quadrilaterals. Among all the bicentric quadrilaterals with a given two circle radii, the one with maximum area is a right kite.

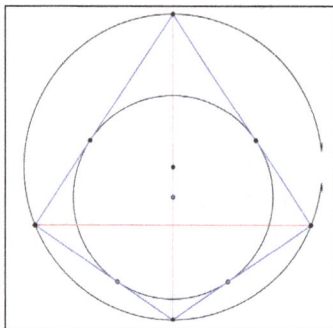

A right kite.

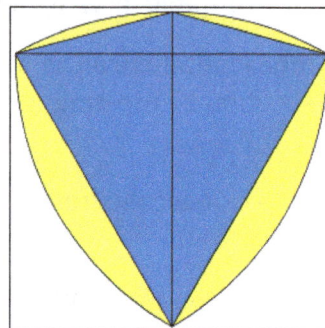

An equidiagonal kite inscribed in a Reuleaux triangle.

There are only eight polygons that can tile the plane in such a way that reflecting any tile across any one of its edges produces another tile; one of them is a right kite, with 60°, 90°, and 120° angles. The tiling that it produces by its reflections is the deltoidal trihexagonal tiling.

Among all quadrilaterals, the shape that has the greatest ratio of its perimeter to its diameter is an equidiagonal kite with angles $\pi/3$, $5\pi/12$, $5\pi/6$, $5\pi/12$. Its four vertices lie at the three corners and one of the side midpoints of the Reuleaux triangle (above to the right).

In non-Euclidean geometry, a Lambert quadrilateral is a right kite with three right angles.

Characterizations

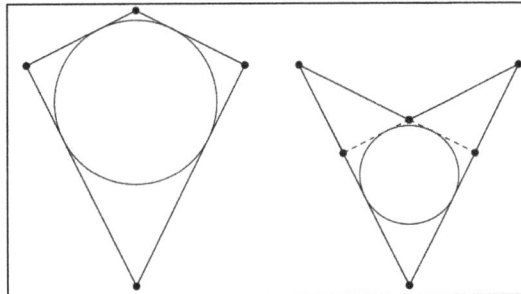

Example convex and concave kites. The concave case is called a dart.

A quadrilateral is a kite if and only if any one of the following conditions is true:

- Two disjoint pairs of adjacent sides are equal (by definition).

- One diagonal is the perpendicular bisector of the other diagonal. (In the concave case it is the extension of one of the diagonals).

- One diagonal is a line of symmetry (it divides the quadrilateral into two congruent triangles that are mirror images of each other).

- One diagonal bisects a pair of opposite angles.

Symmetry

The kites are the quadrilaterals that have an axis of symmetry along one of their diagonals. Any non-self-crossing quadrilateral that has an axis of symmetry must be either a kite (if the axis of symmetry is a diagonal) or an isosceles trapezoid (if the axis of symmetry passes through the midpoints of two sides); these include as special cases the rhombus and the rectangle respectively, which have two axes of symmetry each, and the square which is both a kite and an isosceles trapezoid and has four axes of symmetry. If crossings are allowed, the list of quadrilaterals with axes of symmetry must be expanded to also include the antiparallelograms.

Basic Properties

Every kite is orthodiagonal, meaning that its two diagonals are at right angles to each other. Moreover, one of the two diagonals (the symmetry axis) is the perpendicular bisector of the other, and is also the angle bisector of the two angles it meets.

One of the two diagonals of a convex kite divides it into two isosceles triangles; the other (the axis of symmetry) divides the kite into two congruent triangles. The two interior angles of a kite that are on opposite sides of the symmetry axis are equal.

Area

As is true more generally for any orthodiagonal quadrilateral, the area A of a kite may be calculated as half the product of the lengths of the diagonals p and q:

$$A = \frac{p \cdot q}{2}.$$

Alternatively, if a and b are the lengths of two unequal sides, and θ is the angle between unequal sides, then the area is:

$$A = ab \cdot \sin \theta.$$

Tangent Circles

Every *convex* kite has an inscribed circle; that is, there exists a circle that is tangent to all four sides. Therefore, every convex kite is a tangential quadrilateral. Additionally, if a convex kite is not a rhombus, there is another circle, outside the kite, tangent to the lines that pass through its four sides; therefore, every convex kite that is not a rhombus is an ex-tangential quadrilateral.

For every *concave* kite there exist two circles tangent to all four (possibly extended) sides: one is interior to the kite and touches the two sides opposite from the concave angle, while the other circle is exterior to the kite and touches the kite on the two edges incident to the concave angle.

Dual Properties

Kites and isosceles trapezoids are dual: the polar figure of a kite is an isosceles trapezoid, and vice versa. The side-angle duality of kites and isosceles trapezoids are compared in the table below:

Isosceles trapezoid	Kite
Two pairs of equal adjacent angles	Two pairs of equal adjacent sides
One pair of equal opposite sides	One pair of equal opposite angles
An axis of symmetry through one pair of opposite sides	An axis of symmetry through one pair of opposite angles
Circumscribed circle	Inscribed circle

Tilings and Polyhedra

All kites tile the plane by repeated inversion around the midpoints of their edges, as do more generally all quadrilaterals. A kite with angles $\pi/3$, $\pi/2$, $2\pi/3$, $\pi/2$ can also tile the plane by repeated reflection across its edges; the resulting tessellation, the deltoidal trihexagonal tiling, superposes a tessellation of the plane by regular hexagons and isosceles triangles.

The deltoidal icositetrahedron, deltoidal hexecontahedron, and trapezohedron are polyhedra with congruent kite-shaped facets. There are an infinite number of uniform tilings of the hyperbolic plane by kites, the simplest of which is the deltoidal triheptagonal tiling.

Kites and darts in which the two isosceles triangles forming the kite have apex angles of $2\pi/5$ and $4\pi/5$ represent one of two sets of essential tiles in the Penrose tiling, an aperiodic tiling of the plane discovered by mathematical physicist Roger Penrose.

Face-transitive self-tesselation of the sphere, Euclidean plane, and hyperbolic plane with kites occurs as uniform duals: $\phi_p \bullet_q \phi$ for Coxeter group [p,q], with any set of p,q between 3 and infinity, as this table partially shows up to q=6. When p=q, the kites become rhombi.

Conditions for when a Tangential Quadrilateral is a Kite

A tangential quadrilateral is a kite if and only if any one of the following conditions is true:

- The area is one half the product of the diagonals.

- The diagonals are perpendicular. (Thus the kites are exactly the quadrilaterals that are both tangential and orthodiagonal).

- The two line segments connecting opposite points of tangency have equal length.

- One pair of opposite tangent lengths have equal length.

- The bimedians have equal length.

- The products of opposite sides are equal.

- The center of the incircle lies on a line of symmetry that is also a diagonal.

If the diagonals in a tangential quadrilateral $ABCD$ intersect at P, and the incircles in triangles ABP, BCP, CDP, DAP have radii r_1, r_2, r_3, and r_4 respectively, then the quadrilateral is a kite if and only if:

$$r_1 + r_3 = r_2 + r_4.$$

If the excircles to the same four triangles opposite the vertex P have radii R_1, R_2, R_3, and R_4 respectively, then the quadrilateral is a kite if and only if:

$$R_1 + R_3 = R_2 + R_4.$$

PARALLELOGRAM

In Euclidean geometry, a parallelogram is a simple (non-self-intersecting) quadrilateral with two pairs of parallel sides. The opposite or facing sides of a parallelogram are of equal length and the opposite angles of a parallelogram are of equal measure. The congruence of opposite sides and opposite angles is a direct consequence of the Euclidean parallel postulate and neither condition can be proven without appealing to the Euclidean parallel postulate or one of its equivalent formulations.

The three-dimensional counterpart of a parallelogram is a parallelepiped.

Special Cases

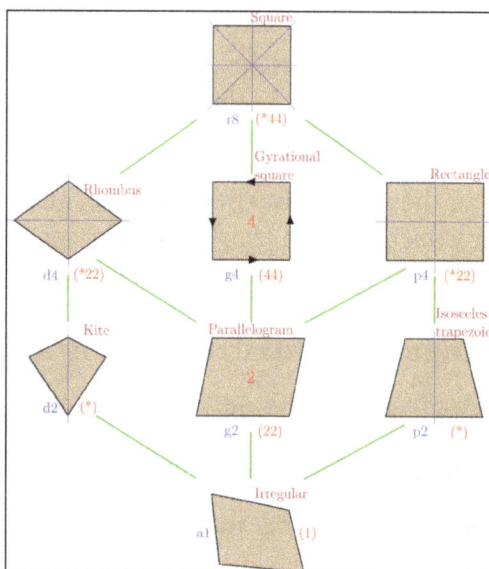

Quadrilaterals by symmetry.

- Rhomboid – A quadrilateral whose opposite sides are parallel and adjacent sides are unequal, and whose angles are not right angles.

- Rectangle – A parallelogram with four angles of equal size (right angles).

- Rhombus – A parallelogram with four sides of equal length.

- Square – A parallelogram with four sides of equal length and angles of equal size (right angles).

Characterizations

A simple (non-self-intersecting) quadrilateral is a parallelogram if and only if any one of the following statements is true:

- Two pairs of opposite sides are parallel (by definition).

- Two pairs of opposite sides are equal in length.

- Two pairs of opposite angles are equal in measure.

- The diagonals bisect each other.

- One pair of opposite sides is parallel and equal in length.

- Adjacent angles are supplementary.

- Each diagonal divides the quadrilateral into two congruent triangles.

- The sum of the squares of the sides equals the sum of the squares of the diagonals. (This is the parallelogram law).

- It has rotational symmetry of order 2.

- The sum of the distances from any interior point to the sides is independent of the location of the point. (This is an extension of Viviani's theorem).
- There is a point X in the plane of the quadrilateral with the property that every straight line through X divides the quadrilateral into two regions of equal area.

Thus all parallelograms have all the properties listed above, and conversely, if just one of these statements is true in a simple quadrilateral, then it is a parallelogram.

Other Properties

- Opposite sides of a parallelogram are parallel (by definition) and so will never intersect.
- The area of a parallelogram is twice the area of a triangle created by one of its diagonals.
- The area of a parallelogram is also equal to the magnitude of the vector cross product of two adjacent sides.
- Any line through the midpoint of a parallelogram bisects the area.
- Any non-degenerate affine transformation takes a parallelogram to another parallelogram.
- A parallelogram has rotational symmetry of order 2 (through 180°) (or order 4 if a square). If it also has exactly two lines of reflectional symmetry then it must be a rhombus or an oblong (a non-square rectangle). If it has four lines of reflectional symmetry, it is a square.
- The perimeter of a parallelogram is $2(a + b)$ where a and b are the lengths of adjacent sides.
- Unlike any other convex polygon, a parallelogram cannot be inscribed in any triangle with less than twice its area.
- The centers of four squares all constructed either internally or externally on the sides of a parallelogram are the vertices of a square.
- If two lines parallel to sides of a parallelogram are constructed concurrent to a diagonal, then the parallelograms formed on opposite sides of that diagonal are equal in area.
- The diagonals of a parallelogram divide it into four triangles of equal area.

Area Formula

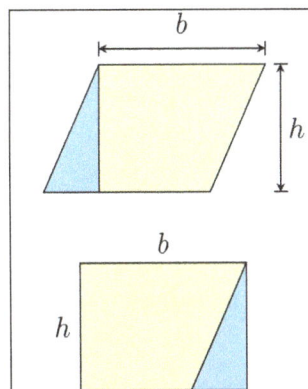

A parallelogram can be rearranged into a rectangle with the same area.

All of the area formulas for general convex quadrilaterals apply to parallelograms. Further formulas are specific to parallelograms:

A parallelogram with base b and height h can be divided into a trapezoid and a right triangle, and rearranged into a rectangle, as shown in the figure to the left. This means that the area of a parallelogram is the same as that of a rectangle with the same base and height:

$$K = bh.$$

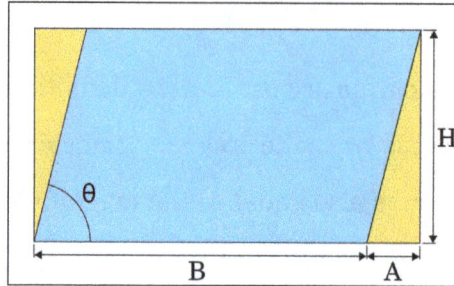

The area of the parallelogram is the area of the blue region, which is the interior of the parallelogram.

The base × height area formula can also be derived using the figure to the right. The area K of the parallelogram to the right (the blue area) is the total area of the rectangle less the area of the two orange triangles. The area of the rectangle is:

$$K_{rect} = (B+A) \times H$$

and the area of a single orange triangle is:

$$K_{tri} = \frac{1}{2} A \times H.$$

Therefore, the area of the parallelogram is:

$$K = K_{rect} - 2 \times K_{tri} = ((B+A) \times H) - (A \times H) = B \times H.$$

Another area formula, for two sides B and C and angle θ, is:

$$K = B \cdot C \cdot \sin \theta.$$

The area of a parallelogram with sides B and C ($B \neq C$) and angle at the intersection of the diagonals is given by:

$$K = \frac{|\tan \gamma|}{2} \cdot \left| B^2 - C^2 \right|.$$

When the parallelogram is specified from the lengths B and C of two adjacent sides together with the length D_1 of either diagonal, then the area can be found from Heron's formula. Specifically it is:

$$K = 2\sqrt{S(S-B)(S-C)(S-D_1)}$$

where $S = (B + C + D_1) / 2$ and the leading factor 2 comes from the fact that the chosen diagonal divides the parallelogram into *two* congruent triangles.

Area in Terms of Cartesian Coordinates of Vertices

Let vectors $a, b \in \mathbb{R}^2$ and let $V = \begin{bmatrix} a_1 & a_2 \\ b_1 & b_2 \end{bmatrix} \in \mathbb{R}^{2 \times 2}$ denote the matrix with elements of a and b. Then the area of the parallelogram generated by a and b is equal to $\det(V) = | a_1 b_2 - a_2 b_1 |$.

Let vectors $| a, b \in \mathbb{R}^n$ and let $V = \begin{bmatrix} a_1 & a_2 & \cdots & a_n \\ b_1 & b_2 & \cdots & b_n \end{bmatrix} \in \mathbb{R}^{2 \times n}$. Then the area of the parallelogram

generated by a and b is equal to $\sqrt{\det(VV^{\mathrm{T}})}$.

Let points $a, b, c \in \mathbb{R}^2$. Then the area of the parallelogram with vertices at a, b and c is equivalent to the absolute value of the determinant of a matrix built using a, b and c as rows with the last column padded using ones as follows:

$$K = \left| \det \begin{bmatrix} a_1 & a_2 & 1 \\ b_1 & b_2 & 1 \\ c_1 & c_2 & 1 \end{bmatrix} \right|.$$

Proof that Diagonals Bisect each other

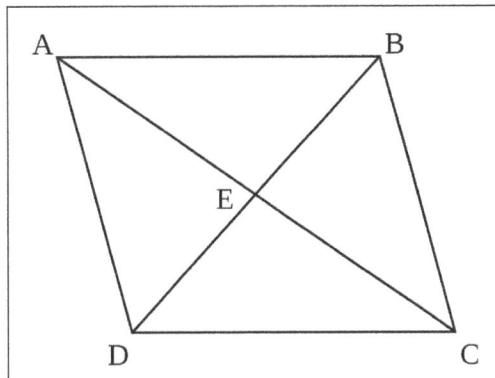

To prove that the diagonals of a parallelogram bisect each other, we will use congruent triangles:

- $\angle ABE \cong \angle CDE$ *(alternate interior angles are equal in measure).*

- $\angle BAE \cong \angle DCE$ *(alternate interior angles are equal in measure).*

(since these are angles that a transversal makes with parallel lines *AB* and *DC*).

Also, side *AB* is equal in length to side *DC*, since opposite sides of a parallelogram are equal in length.

Therefore, triangles ABE and CDE are congruent (ASA postulate, two corresponding angles and the included side).

Therefore,

$$AE = CE$$
$$BE = DE.$$

Since the diagonals AC and BD divide each other into segments of equal length, the diagonals bisect each other.

Separately, since the diagonals AC and BD bisect each other at point E, point E is the midpoint of each diagonal.

Lattice of Parallelograms

Parallelograms can tile the plane by translation. If edges are equal, or angles are right, the symmetry of the lattice is higher. These represent the four Bravais lattices in 2 dimensions.

Lattices				
Form	Square	Rectangle	Rhombus	Parallelogram
System	Square (tetragonal)	Rectangular (orthorhombic)	Centered rectangular (orthorhombic)	Oblique (monoclinic)
Constraints	$\alpha=90°$, a=b	$\alpha=90°$	a=b	None
Symmetry	p4m, [4,4], order 8n	pmm, [∞,2,∞], order 4n		p1, [∞+,2,∞+], order 2n
Form				

Parallelograms Arising from other Figures

Automedian Triangle

An automedian triangle is one whose medians are in the same proportions as its sides (though in a different order). If ABC is an automedian triangle in which vertex A stands opposite the side a, G is the centroid (where the three medians of ABC intersect), and AL is one of the extended medians of ABC with L lying on the circumcircle of ABC, then $BGCL$ is a parallelogram.

Varignon Parallelogram

The midpoints of the sides of an arbitrary quadrilateral are the vertices of a parallelogram, called its Varignon parallelogram. If the quadrilateral is convex or concave (that is, not self-intersecting), then the area of the Varignon parallelogram is half the area of the quadrilateral.

Tangent Parallelogram of An Ellipse

For an ellipse, two diameters are said to be conjugate if and only if the tangent line to the ellipse at an endpoint of one diameter is parallel to the other diameter. Each pair of conjugate diameters of an ellipse has a corresponding tangent parallelogram, sometimes called a bounding parallelogram, formed by the tangent lines to the ellipse at the four endpoints of the conjugate diameters. All tangent parallelograms for a given ellipse have the same area.

It is possible to reconstruct an ellipse from any pair of conjugate diameters, or from any tangent parallelogram.

Faces of a Parallelepiped

A parallelepiped is a three-dimensional figure whose six faces are parallelograms.

TRAPEZIUM

A trapezium, also known as a trapezoid, is a quadrilateral in which a pair of sides are parallel, but the other pair of opposite sides are non-parallel. The area of a trapezium is computed with the following formula:

$$Area = \frac{1}{2} \times Sum\,of\,parallel\,sides \times Distance\,between\,them.$$

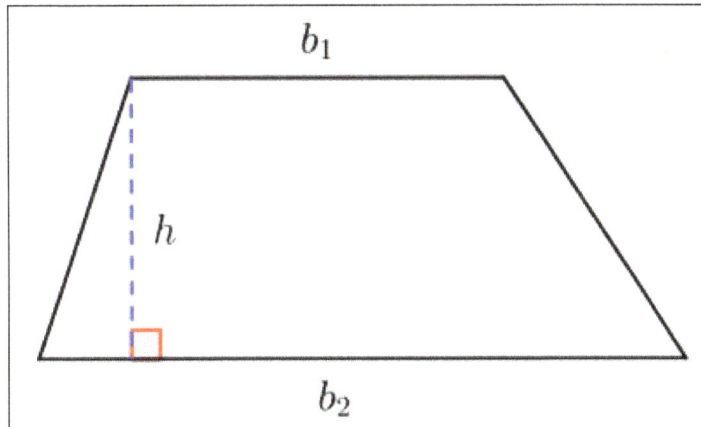

The parallel sides are called the bases of the trapezium. Let $b_1\,and\,b_2$ be the lengths of these bases. The distance between the bases is called the height of the trapezium. Let hhh be this height. Then this formula becomes:

$$Area = \frac{1}{2}(b_1 + b_2)h$$

Proof

Given a trapezium, let $b_1\,and\,b_2$ be the lengths of the bases, and let h be the height. Draw a segment parallel to the bases that is halfway between the bases. This divides the trapezium into two trapeziums, each with the same height of $\frac{1}{2}h.$

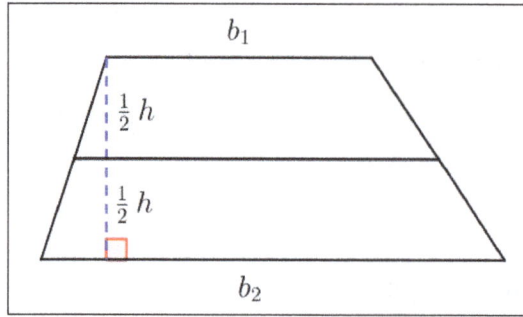

Labeling the angles of these trapeziums:

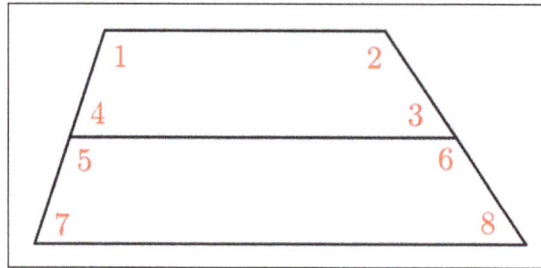

The following congruences and identities due to the fact that the bases are parallel:

$$m\angle 4 + m\angle 5 = 180°$$

$$m\angle 1 + m\angle 7 = 180°$$

$$\angle 2 \cong \angle 6$$

$$\angle 3 \cong \angle 8$$

Now rotate the top trapezoid and place it adjacent to the bottom trapezoid in the following way:

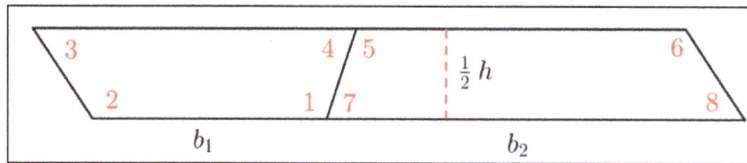

Due to the aforementioned congruences and identities, this shape is a parallelogram. The length of its base is $(b_1 + b_2)$, and its height is $\frac{1}{2}h$. This parallelogram has the same area as the trapezoid, so the area of the trapezoid is:

$$Area = \frac{1}{2}(b_1 + b_2)h.$$

SPHERE

Sphere is the set of all points in three-dimensional space lying the same distance (the radius) from a given point (the centre), or the result of rotating a circle about one of its diameters. The components and properties of a sphere are analogous to those of a circle. A diameter is any line segment connecting two points of a sphere and passing through its centre. The circumference is the length of any great circle, the intersection of the sphere with any plane passing through its centre. A meridian is any great circle passing through a point designated a pole. A geodesic, the shortest distance between any two points on a sphere, is an arc of the great circle through the two points. The formula for determining a sphere's surface area is $4\pi r2$; its volume is determined by $(4/3)\pi r3$. The study of spheres is basic to terrestrial geography and is one of the principal areas of Euclidean geometry and elliptic geometry.

CONE

Cones are pyramid-like structures with a circular base. It is a three-dimensional shape and is a smooth base that tapers at a point called vertex. While studying the volume and surface area of cones, we generally consider a right circular cone which has a circular base and the axis passes through the center of this base making a right angle from the base.

The other kind of cone, however, is the oblique cone in which the axis is non-perpendicular to the base. Cones can be made from circular sheets, rolled inwards. On joining the two sides the shape which we get is the cone.

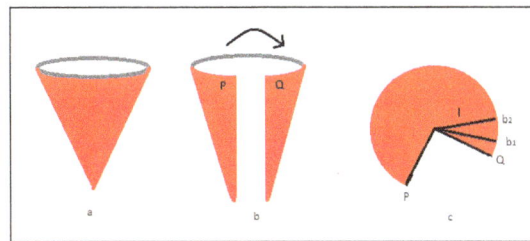

cone

Surface Area of a Cone

The area of each triangle = $1/2\times$base of triangle$\times$l. The circle (c) is a sum of many triangles. So the area of the entire circle is the sum of the areas of all triangles = $1/2\ b_1l + 1/2\ b_2l + 1/2\ b_3l +\ldots\ldots$ = $1/2\ l\ (b_1 + b_2 + b_3 + \ldots) = 1/2 \times l \times$ (length of the whole curved boundary). Now the entire curved portion makes the perimeter of the base of cones. The circumference of the base of the cone = $2\pi r$, here r is the base of the radius of the circular base.

Cone

So the curved surface area = 1/2 *l* $2\pi r$ = πrl. r is the radius of the base circle and l is its slant height. We already know that cones constitute the right-angled triangle. In a right-angled triangle, Pythagoras Theorem helps us find out the length of the slant side, with the help of formula: $l^2 = r^2 + h^2$.

Therefore, Slant Height (l) = $\sqrt{r^2 + h^2}$. The base of the cone is closed with the help of the circle. The area of the circular base is πr^2. So, the total surface area = $\pi rl + \pi r^2$. Therefore,

The Surface Area of a Cone = $\pi r (l+r)$.

The Volume of Right Circular Cones

Cones are 3-dimensional triangles with a circular base. According to the structure, the volume of a cone is assumed to be 1/3 of a cylinder with the same sized circular base, height, and length. The volume of a cylinder is $\pi r^2 h$. And hence that of the cone is 1/3 $\pi r^2 h$, where r is the radius of the circular base while h is the height of the cone. Therefore,

The Volume of a Right Circular Cone = 1/3 $\pi r^2 h$

The volume of a right circular is 9856 cm³. If the diameter of the base is 28cm, find the cone's height, slant height and curved surface area?

Solution: Volume (V) = 9856 cm³

Diameter of the circular base = 28 cm.

Radius of the circular base= d/2 = 28/2= 14 cm

1/3 $\pi r^2 h$ = 9856 cm³ = 1/3 × 22/7 × 14 × 14 × h = 9856.

Also, h= 9856 × 3 × 7 /14 × 14 × 22 = 48 cm

Height of the cone = 48 cm. Slant height (l) = $\sqrt{r^2 + h^2}$ = $\sqrt{14^2} + \sqrt{48^2}$

Slant height of the cone = 50 cm.

Curved Surface area = πrl = 22/7 × 14 × 50.

Curved surface area of the cone= 2200 cm².

CUBOID

In geometry, a cuboid is a convex polyhedron bounded by six quadrilateral faces, whose polyhedral graph is the same as that of a cube. While mathematical literature refers to any such polyhedron as a cuboid, other sources use "cuboid" to refer to a shape of this type in which each of the faces is a rectangle (and so each pair of adjacent faces meets in a right angle); this more restrictive type of cuboid is also known as a rectangular cuboid, right cuboid, rectangular box, rectangular hexahedron, right rectangular prism, or rectangular parallelepiped.

General Cuboids

By Euler's formula the numbers of faces F, of vertices V, and of edges E of any convex polyhedron are related by the formula $F + V = E + 2$. In the case of a cuboid this gives $6 + 8 = 12 + 2$; that is, like a cube, a cuboid has 6 faces, 8 vertices, and 12 edges. Along with the rectangular cuboids, any parallelepiped is a cuboid of this type, as is a square frustum (the shape formed by truncation of the apex of a square pyramid).

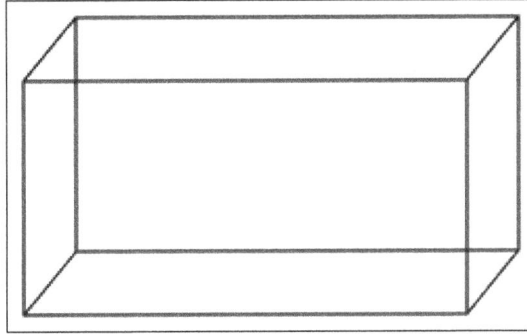

Rectangular Cuboid

In a rectangular cuboid, all angles are right angles, and opposite faces of a cuboid are equal. By definition this makes it a right rectangular prism, and the terms *rectangular parallelepiped* or *orthogonal parallelepiped* are also used to designate this polyhedron. The terms "rectangular prism" and "oblong prism", however, are ambiguous, since they do not specify all angles.

The square cuboid, square box, or right square prism (also ambiguously called *square prism*) is a special case of the cuboid in which at least two faces are squares. It has Schläfli symbol $\{4\} \times \{\ \}$, and its symmetry is doubled from [2,2] to [4,2], order 16.

The cube is a special case of the square cuboid in which all six faces are squares. It has Schläfli symbol $\{4,3\}$, and its symmetry is raised from [2,2], to [4,3], order 48.

If the dimensions of a rectangular cuboid are a, b and c, then its volume is abc and its surface area is $2(ab + ac + bc)$.

The length of the space diagonal is:

$$d = \sqrt{a^2 + b^2 + c^2}.$$

Cuboid shapes are often used for boxes, cupboards, rooms, buildings, containers, cabinets, books, a sturdy computer chassis, printing devices, electronic calling touchscreen devices, washing and drying machines, etc. Cuboids are among those solids that can tessellate 3-dimensional space. The shape is fairly versatile in being able to contain multiple smaller cuboids, e.g. sugar cubes in a box, boxes in a cupboard, cupboards in a room, and rooms in a building.

A cuboid with integer edges as well as integer face diagonals is called an Euler brick, for example with sides 44, 117 and 240. A perfect cuboid is an Euler brick whose space diagonal is also an integer. It is currently unknown whether a perfect cuboid actually exists.

Nets

The number of different nets for a simple cube is 11, however this number increases significantly to 54 for a rectangular cuboid of 3 different lengths.

CUBE

In Euclidean geometry, cube is a regular solid with six square faces; that is, a regular hexahedron.

Since the volume of a cube is expressed, in terms of an edge e, as e^3, in arithmetic and algebra the third power of a quantity is called the cube of that quantity. That is, 3^3, or 27, is the cube of 3, and x^3 is the cube of x. A number of which a given number is the cube is called the cube root of the latter number; that is, since 27 is the cube of 3, 3 is the cube root of 27—symbolically, $3 = \sqrt[3]{27}$. A number that is not a cube is also said to have a cube root, the value being expressed approximately; that is, 4 is not a cube, but the cube root of 4 is expressed as $\sqrt[3]{4}$, the approximate value being 1.587.

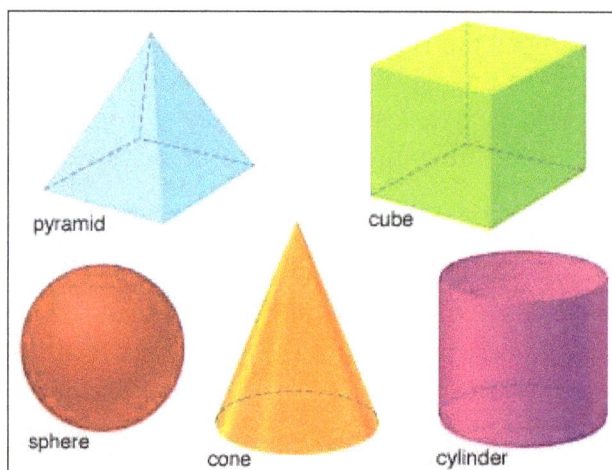

In Greek geometry the duplication of the cube was one of the most famous of the unsolved problems. It required the construction of a cube that should have twice the volume of a given cube. This proved to be impossible by the aid of the straight edge and compasses alone, but the Greeks were able to effect the construction by the use of higher curves, notably by the cissoid of Diocles. Hippocrates showed that the problem reduced to that of finding two mean proportionals between a line segment and its double—that is, algebraically, to that of finding x and y in the proportion a:x = x:y = y:2a, from which $x^3 = 2a^3$, and hence the cube with x as an edge has twice the volume of one with a as an edge.

Cube Formula

The two major formulas are for surface area and volume of a solid shape.

- Surface Area of Cube = $6a^2$ in a square unit.

- Volume of the cube = a^3 in cubic units.

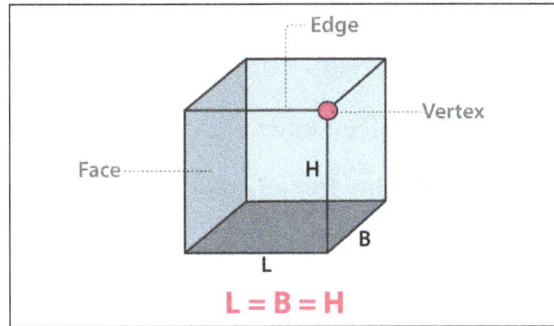

Properties

- It has faces in a square shape.

- All the faces or sides have equal dimensions.

- The plane angles of the cube are the right angle.

- Each of the faces meets the other four faces.

- Each of the vertices meets the three faces and three edges.

- The edges opposite to each other in a cube are parallel.

- If a is the length of the side, then.

- Length of Diagonal of Face of the Cube = $\sqrt{2}$ a.

- Length of Diagonal of Cube = $\sqrt{3}$ a.

Cube Facts

- A cube is a three dimensional shape that features all right angles and a height, width and depth that are all equal.

- A cube has 6 square faces.

- A cube has 8 points (vertices).

- A cube has 12 edges.

- Things that are shaped like a cube are often referred to as 'cubic'.

- A cube is a special geometric shape that falls into a number of groups including platonic solids and regular hexahedrons.

- The surface area of a cube can be found with the following formula (where a = the length of an edge): Surface area = $6a^2$.

- In other words: Surface area = 6 × edge × edge.

- The volume of a cube can be found with the following formula (where a = the length of an edge): Volume = a³.

- In other words: Volume = edge × edge × edge.

- A cube has the largest volume of all cuboids with a certain surface area.

- Most dice are cube shaped, featuring the numbers 1 to 6 on the different faces.

- 11 different 'nets' can be made by folding out the 6 square faces of a cube.

- A square is in many ways like a cube, only in two dimensions rather than three.

PRISM

In geometry, a prism is a polyhedron comprising an *n*-sided polygonal base, a second base which is a translated copy (rigidly moved without rotation) of the first, and *n* other faces (necessarily all parallelograms) joining corresponding sides of the two bases. All cross-sections parallel to the bases are translations of the bases. Prisms are named for their bases, so a prism with a pentagonal base is called a pentagonal prism. The prisms are a subclass of the prismatoids.

General, Right and Uniform Prisms

A right prism is a prism in which the joining edges and faces are perpendicular to the base faces. This applies if the joining faces are rectangular. If the joining edges and faces are not perpendicular to the base faces, it is called an oblique prism.

For example a parallelepiped is an *oblique prism* of which the base is a parallelogram, or equivalently a polyhedron with six faces which are all parallelograms.

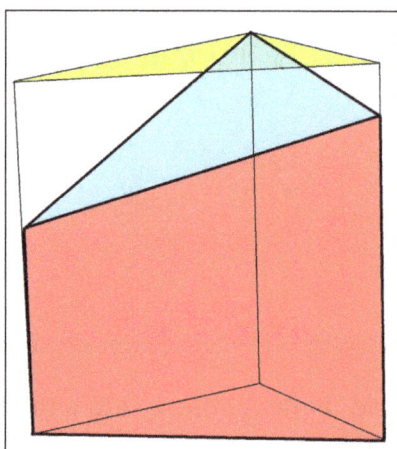

A truncated triangular prism with its top face truncated at an oblique angle.

A truncated prism is a prism with nonparallel top and bottom faces.

Some texts may apply the term rectangular prism or square prism to both a right rectangular-sided prism and a right square-sided prism. A *right p-gonal prism* with rectangular sides has a Schläfli symbol { } × {p}.

A right rectangular prism is also called a *cuboid*, or informally a *rectangular box*. A right square prism is simply a *square box*, and may also be called a *square cuboid*. A *right rectangular prism* has Schläfli symbol { }×{ }×{ }.

An *n*-prism, having regular polygon ends and rectangular sides, approaches a cylindrical solid as *n* approaches infinity.

The term uniform prism or *semiregular prism* can be used for a *right prism* with square sides, since such prisms are in the set of uniform polyhedra. A *uniform p-gonal prism* has a Schläfli symbol t{2,p}. Right prisms with regular bases and equal edge lengths form one of the two infinite series of semiregular polyhedra, the other series being the antiprisms.

The dual of a *right prism* is a bipyramid.

Volume

The volume of a prism is the product of the area of the base and the distance between the two base faces, or the height (in the case of a non-right prism, note that this means the perpendicular distance).

The volume is therefore:

$$V = Bh$$

where B is the base area and h is the height. The volume of a prism whose base is a regular n-sided polygon with side length s is therefore:

$$V = \frac{n}{4} hs^2 \cot(\frac{\pi}{n})$$

Surface Area

The surface area of a right prism is:

$$2B + Ph$$

where B is the area of the base, h the height, and P the base perimeter.

The surface area of a right prism whose base is a regular n-sided polygon with side length s and height h is therefore:

$$A = \frac{n}{2} s^2 \cot \frac{\pi}{n} + nsh$$

Symmetry

The symmetry group of a right n-sided prism with regular base is D_{nh} of order $4n$, except in the case of a cube, which has the larger symmetry group O_h of order 48, which has three versions of D_{4h} as subgroups. The rotation group is D_n of order $2n$, except in the case of a cube, which has the larger symmetry group O of order 24, which has three versions of D_4 as subgroups.

The symmetry group D_{nh} contains inversion iff n is even.

The hosohedra and dihedra also possess dihedral symmetry, and a n-gonal prism can be constructed via the geometrical truncation of a n-gonal hosohedron, as well as through the cantellation or expansion of a n-gonal dihedron.

Prismatic Polytope

A *prismatic polytope* is a higher-dimensional generalization of a prism. An n-dimensional prismatic polytope is constructed from two $(n - 1)$-dimensional polytopes, translated into the next dimension.

The prismatic n-polytope elements are doubled from the $(n - 1)$-polytope elements and then creating new elements from the next lower element.

Take an n-polytope with f_i i-face elements ($i = 0, ..., n$). Its $(n + 1)$-polytope prism will have $2f_i + f_{i-1}$ i-face elements. (With $f_{-1} = 0, f_n = 1$.)

By dimension:

- Take a polygon with n vertices, n edges. Its prism has $2n$ vertices, $3n$ edges, and $2 + n$ faces.

- Take a polyhedron with v vertices, e edges, and f faces. Its prism has $2v$ vertices, $2e + v$ edges, $2f + e$ faces, and $2 + f$ cells.

- Take a polychoron with v vertices, e edges, f faces and c cells. Its prism has $2v$ vertices, $2e + v$ edges, $2f + e$ faces, and $2c + f$ cells, and $2 + c$ hypercells.

Uniform Prismatic Polytope

A regular n-polytope represented by Schläfli symbol $\{p, q, ..., t\}$ can form a uniform prismatic $(n + 1)$-polytope represented by a Cartesian product of two Schläfli symbols: $\{p, q, ..., t\}×\{\}$.

By dimension:

- A 0-polytopic prism is a line segment, represented by an empty Schläfli symbol $\{\}$.

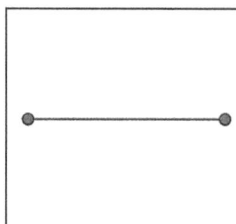

- A 1-polytopic prism is a rectangle, made from 2 translated line segments. It is represented as the product Schläfli symbol $\{\}×\{\}$. If it is square, symmetry can be reduced: $\{\}×\{\} = \{4\}$.

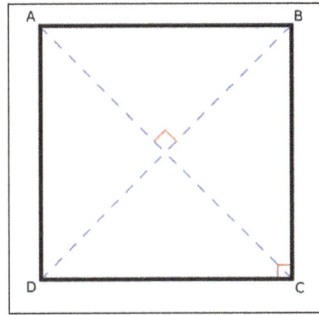

- ◦ Example: Square, {}×{}, two parallel line segments, connected by two line segment sides.

- A polygonal prism is a 3-dimensional prism made from two translated polygons connected by rectangles. A regular polygon {p} can construct a uniform n-gonal prism represented by the product {p}×{}. If p = 4, with square sides symmetry it becomes a cube: {4}×{} = {4, 3}.

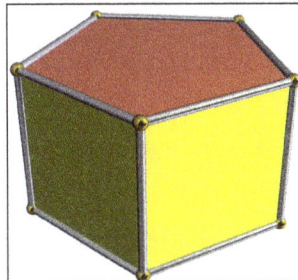

- ◦ Example: Pentagonal prism, {5}×{}, two parallel pentagons connected by 5 rectangular *sides*.

- A polyhedral prism is a 4-dimensional prism made from two translated polyhedra connected by 3-dimensional prism cells. A regular polyhedron {p, q} can construct the uniform polychoric prism, represented by the product {p, q}×{}. If the polyhedron is a cube, and the sides are cubes, it becomes a tesseract: {4, 3}×{} = {4, 3, 3}.

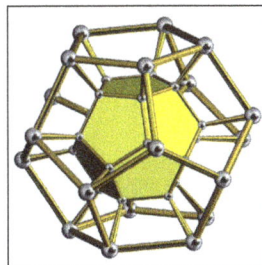

- ◦ Example: Dodecahedral prism, {5, 3}×{}, two parallel dodecahedra connected by 12 pentagonal prism *sides*.

Higher order prismatic polytopes also exist as cartesian products of any two polytopes. The dimension of a polytope is the product of the dimensions of the elements. The first example of these exist in 4-dimensional space are called duoprisms as the product of two polygons. Regular duoprisms are represented as {p}×{q}.

Twisted Prism

A twisted prism is a nonconvex prism polyhedron constructed by a uniform q-prism with the side faces bisected on the square diagonal, and twisting the top, usually by $\frac{\pi}{q}$ radians ($\frac{180}{q}$ degrees) in the same direction, causing side triangles to be concave.

A twisted prism cannot be dissected into tetrahedra without adding new vertices. The smallest case, triangular form, is called a Schönhardt polyhedron.

A *twisted prism* is topologically identical to the antiprism, but has half the symmetry: D_n, $[n,2]^+$, order $2n$. It can be seen as a convex antiprism, with tetrahedra removed between pairs of triangles.

Frustum

A frustum is topologically identical to a prism, with trapezoid lateral faces and different sized top and bottom polygons.

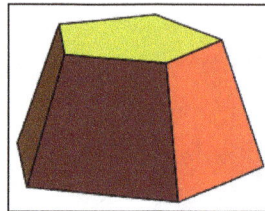

Pentagonal frustum

Star Prism

A star prism is a nonconvex polyhedron constructed by two identical star polygon faces on the top and bottom, being parallel and offset by a distance and connected by rectangular faces. A *uniform star prism* will have Schläfli symbol $\{p/q\} \times \{\ \}$, with p rectangle and $2\ \{p/q\}$ faces. It is topologically identical to a p-gonal prism.

Examples	
$\{\ \}\times\{\ \}_{180}\times\{\ \}$	$t_a\{3\}\times\{\ \}$
D_{2h}, order 8	D_{3h}, order 12

Crossed Prism

A crossed prism is a nonconvex polyhedron constructed from a prism, where the base vertices are inverted around the center (or rotated 180°). This transforms the side rectangular faces into crossed rectangles. For a regular polygon base, the appearance is an p-gonal hour glass, with all vertical edges passing through a single center, but no vertex is there. It is topologically identical to a p-gonal prism.

Examples	
$\{\}\times\{\}_{180}\times\{\}_{180}$	$t_a\{3\}\times\{\}_{180}$
D_{2h}, order 8	D_{3d}, order 12

Toroidal Prisms

Examples	
D_4h, order 16	D_{6h}, order 24
v=8, e=16, f=8	v=12, e=24, f=12

A toroidal prism is a nonconvex polyhedron is like a *crossed prism* except instead of having base and top polygons, simple rectangular side faces are added to close the polyhedron. This can only be done for even-sided base polygons. These are topological tori, with Euler characteristic of zero. The topological polyhedral net can be cut from two rows of a square tiling, with vertex figure. A n-gonal toroidal prism has $2n$ vertices and faces, and $4n$ edges and is topologically self-dual.

References

- Alsina, claudi; nelsen, roger b. (2009), when less is more: visualizing basic inequalities, the dolciani mathematical expositions, 36, mathematical association of america, washington, dc, isbn 978-0-88385-342-9, mr 2498836

- What-is-a-scalene-triangle-definition-properties-examples, lesson, academy: study.com, Retrieved 10 April, 2019

- Gottschau, marinus; haverkort, herman; matzke, kilian (2018), "reptilings and space-filling curves for acute triangles", discrete & computational geometry, 60 (1): 170–199, arxiv:1603.01382, doi:10.1007/s00454-017-9953-0

- Sphere, science: britannica.com, Retrieved 17 May, 2019

- padovan, richard (2002), towards universality: le corbusier, mies, and de stijl, psychology press, p. 128, isbn 9780415259620

- Rectangle, maths: byjus.com, Retrieved 14 July, 2019

- darling, david (2004), the universal book of mathematics: from abracadabra to zeno's paradoxes, john wiley & sons, p. 260, isbn 9780471667001

- Cone, surface-areas-and-volumes, maths, guides: toppr.com, Retrieved 31 March, 2019

- kirby, matthew; umble, ronald (2011), "edge tessellations and stamp folding puzzles", mathematics magazine, 84 (4): 283–289, arxiv:0908.3257, doi:10.4169/math.mag.84.4.283, mr 2843659

- Area-of-a-trapezium: brilliant.org, Retrieved 30 January, 2019

4

Theorems in Geometry

A hypothesis which is assumed and proved by a chain of reasoning is called a theorem. Geometry consists of several theorems such as Ptolemy's theorem, Brahmagupta's formula, Cauchy's theorem, Finsler–Hadwiger theorem, etc. This chapter has been carefully written to provide an easy understanding of these various theorems of geometry.

PTOLEMY'S THEOREM

Ptolemy's Theorem states that, in a cyclic quadrilateral, the product of the diagonals is equal to the sum the products of the opposite sides.

In the diagram below, Ptolemy's Theorem claims:

$$AC \cdot DB = DC \cdot AB + DA \cdot BC$$

Consider the diagram on the below:

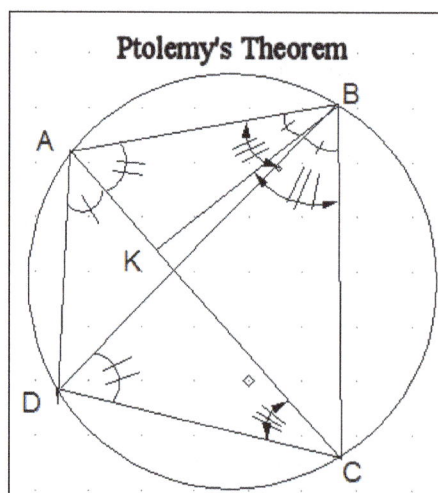

Ptolemy's Theorem

We have drawn a quadrilateral ABCD inside a circle, and constructed the angle ABK so it is equal to angle DBC.

The following facts about the angles in the same segment:

- $\angle ABD = \angle ACD$ (The chord AD subtends equal angles in the same segment.) The angles are marked with one stroke in the diagram.

- $\angle CBD = \angle CAB$ (The chord BC subtends equal angles in the same segment.) The angles are marked with two strokes in the diagram.

- $\angle DBC = \angle DAC$ (The chord DC subtends equal angles in the same segment.) The angles are marked with three strokes in the diagram.

And the following about the named triangles:

- Triangles ABK and DBC are similar, because two of the angles in each are equal (as marked in the diagram), and the third angles are equal because of the angle sum of a triangle.

- The triangles ABC and KBC are similar, because the angles are equal (as marked in the diagram).

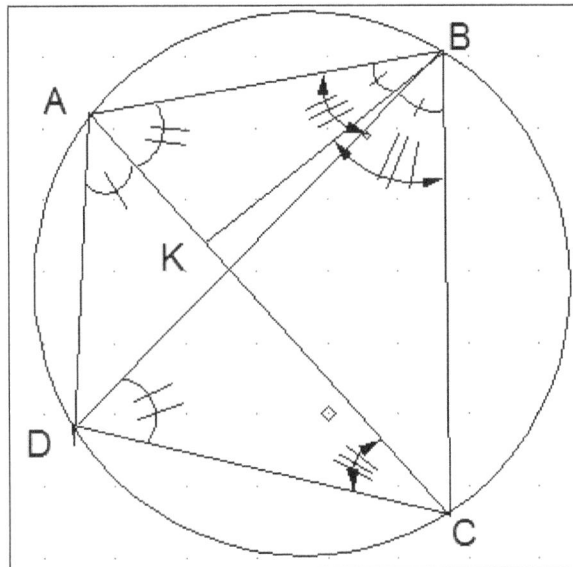

ptolemy's Theorem.

The diagram is repeated on the right to minimize scrolling.

From the triangles ABK and DBC we have:

$$AK \cdot DB = DC \cdot AB$$

From the triangles ABC and KBC, we have:

$$CK \cdot DB = DA \cdot BC$$

By adding (above equations) (and taking out the common factor DB) we have:

$$(AK + CK) \cdot DB = DC \cdot AB + DA \cdot BC$$

Because AK and CK are the parts of AC, we have Ptolemy's Theorem:

$$AC \cdot DB = DC \cdot AB + DA \cdot BC$$

BRAHMAGUPTA'S FORMULA

Brahmagupta's formula provides the area A of a cyclic quadrilateral (i.e., a simple quadrilateral that is inscribed in a circle) with sides of length a, b, c, and d.

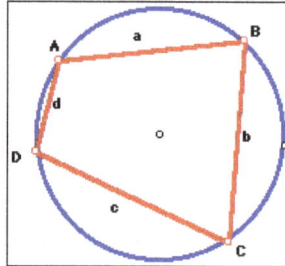

$$A = \sqrt{(s-a)(s-b)(s-c)(s-d)}$$

where s is the semiperimeter:

$$s = \frac{a+b+c+d}{2}$$

If ABCD is a rectangle the formula is clear. Consider the chord AC:

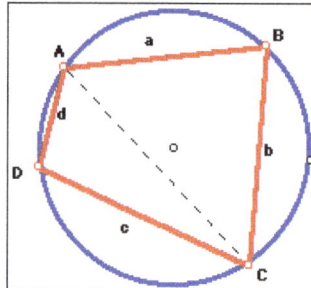

The angle that subtends a chord has measure that is half the measure of the intercepted arc. But the chord AC is simultaneously subtended by the angle at B and by the angle at D. There for the sum of these angles is 180 degrees. Opposite angles of a cyclic quadrilateral are supplemental.

Assume the quadrilateral is not a rectangle. WNLOG, extend AB and CD until they meet at P.

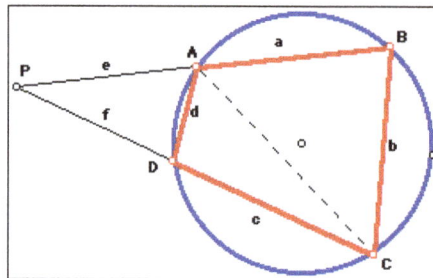

Label the extensions outside the circle e and f.

Now, triangles PBC and PDA are similar. Further, the ratio of similarity is d/b. Therefore the ratio of similarity of their areas is the square of this ratio, or,

$$Area\,of\,triangle\,PDA = \frac{d^2}{b^2}(Area\,of\,triangle\,PBC)$$

Now the area of the quadrilateral ABCD is the area of the larger triangle PBC less the area of the smaller triangle PDA. If A is the area of the quadrilateral and T is the area of triangle PBC:

$$A = T - \frac{d^2}{d^2}T = \left(1 - \frac{d^2}{b^2}\right)T = \frac{b^2 - d^2}{b^2}T$$

or,

$$A = \frac{b^2 - d^2}{b^2}T$$

Now,

1. Heron's formula can be used to express the area of triangle PBC.

2. The similarity of triangles PBC and PAD can be use to effect various (but tedious) substitutions.

With appropriate perseverance and and algebraic substitiutions/simplifications, Brahmagupta's theorem can be derived.

Extensions

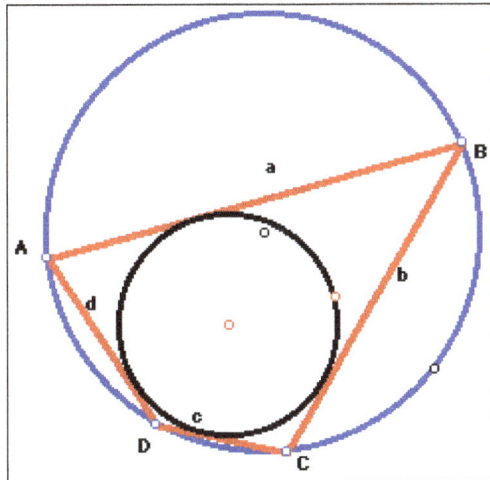

1. If ABCD is a quadrilateral with sides of length a, b, c, and d, such that ABCD is both cyclic and has a circle inscribed in it, then use Brahmagupta's formula to show that the area of the quadrilateral is:

$$A = \sqrt{abcd}$$

2. Consider Brahmagupta's formula as one side, say the one of length d wnlog, varies and approaches zero in length.

3. Use Brahmagupta's formula to develop equations for the length of the two diagonals of the quadrilateral.

CAUCHY'S THEOREM

Cauchy's theorem is a theorem in geometry, named after Augustin Cauchy. It states that convex polytopes in three dimensions with congruent corresponding faces must be congruent to each other. That is, any polyhedral net formed by unfolding the faces of the polyhedron onto a flat surface, together with gluing instructions describing which faces should be connected to each other, uniquely determines the shape of the original polyhedron. For instance, if six squares are connected in the pattern of a cube, then they must form a cube: there is no convex polyhedron with six square faces connected in the same way that does not have the same shape.

This is a fundamental result in rigidity theory: one consequence of the theorem is that, if one makes a physical model of a convex polyhedron by connecting together rigid plates for each of the polyhedron faces with flexible hinges along the polyhedron edges, then this ensemble of plates and hinges will necessarily form a rigid structure.

Let P and Q be *combinatorially equivalent* 3-dimensional convex polytopes; that is, they are convex polytopes with isomorphic face lattices. Suppose further that each pair of corresponding faces from P and Q are congruent to each other, i.e. equal up to a rigid motion. Then P and Q are themselves congruent.

To see that convexity is necessary, consider a regular icosahedron. One can "push in" a vertex to create a nonconvex polyhedron that is still combinatorially equivalent to the regular icosahedron. Another way to see it, is to take the pentagonal pyramid around a vertex, and reflect it with respect to its base.

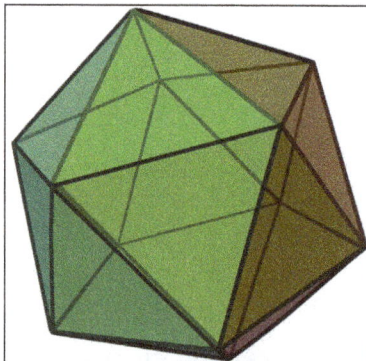
Convex regular icosahedron.

- The result does not hold on a plane or for non-convex polyhedra in R³: there exist non-convex flexible polyhedra that have one or more degrees of freedom of movement that preserve the shapes of their faces. In particular, the Bricard octahedra are self-intersecting flexible surfaces discovered by a French mathematician Raoul Bricard in 1897. The *Connelly sphere*, a flexible non-convex polyhedron homeomorphic to a 2-sphere, was discovered by Robert Connelly in 1977.

- Although originally proven by Cauchy in three dimensions, the theorem was extended to dimensions higher than 3 by Alexandrov.

- Cauchy's rigidity theorem is a corollary from Cauchy's theorem stating that a convex polytope cannot be deformed so that its faces remain rigid.

- In 1974 Herman Gluck showed that in a certain precise sense *almost all* simply connected closed surfaces are rigid.

- Dehn's rigidity theorem is an extension of the Cauchy rigidity theorem to infinitesimal rigidity. This result was obtained by Dehn in 1916.

- Alexandrov's uniqueness theorem is a result by Alexandrov, generalizing Cauchy's theorem by showing that convex polyhedra are uniquely described by the metric spaces of geodesics on their surface. The analogous uniqueness theorem for smooth surfaces was proved by Cohn-Vossen in 1927. Pogorelov's uniqueness theorem is a result by Pogorelov generalizing both of these results and applying to general convex surfaces.

FINSLER–HADWIGER THEOREM

The Finsler–Hadwiger theorem is statement in Euclidean plane geometry that describes a third square derived from any two squares that share a vertex. The theorem is named after Paul Finsler and Hugo Hadwiger, who published it in 1937 as part of the same paper in which they published the Hadwiger–Finsler inequality relating the side lengths and area of a triangle.

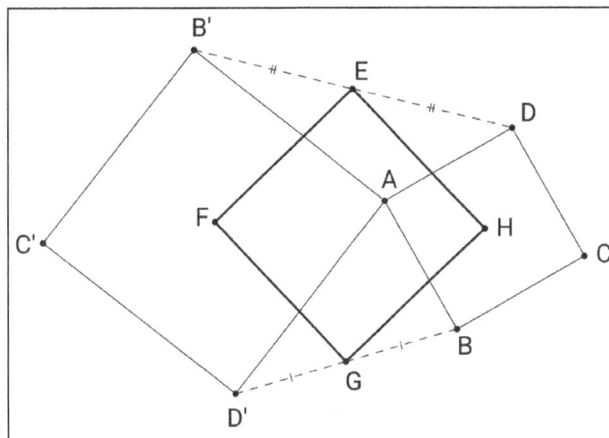

To state the theorem, suppose that ABCD and AB'C'D' are two squares with common vertex A. Let E and G be the midpoints of B'D and D'B respectively, and let F and H be the centers of the two squares. Then the theorem states that the quadrilateral EFGH is a square as well.

The square EFGH is called the Finsler–Hadwiger square of the two given squares.

Application

Repeated application of the Finsler–Hadwiger theorem can be used to prove Van Aubel's theorem, on the congruence and perpendicularity of segments through centers of four squares constructed on the sides of an arbitrary quadrilateral. Each pair of consecutive squares forms an instance of the theorem, and the two pairs of opposite Finsler–Hadwiger squares of those instances form another two instances of the theorem, having the same derived square.

FIVE POINTS DETERMINE A CONIC

In Euclidean and projective geometry, just as two (distinct) points determine a line (a degree-1 plane curve), five points determine a conic (a degree-2 plane curve). There are additional subtleties for conics that do not exist for lines, and thus the statement and its proof for conics are both more technical than for lines.

Formally, given any five points in the plane in general linear position, meaning no three collinear, there is a unique conic passing through them, which will be non-degenerate; this is true over both the Euclidean plane and any pappian projective plane. Indeed, given any five points there is a conic passing through them, but if three of the points are collinear the conic will be degenerate (reducible, because it contains a line), and may not be unique.

This result can be proven numerous different ways; the dimension counting argument is most direct, and generalizes to higher degree, while other proofs are special to conics.

Dimension Counting

Intuitively, passing through five points in general linear position specifies five independent linear constraints on the (projective) linear space of conics, and hence specifies a unique conic, though this brief statement ignores subtleties.

More precisely, this is seen as follows:

- Conics correspond to points in the five-dimensional projective space P^5.

- Requiring a conic to pass through a point imposes a linear condition on the coordinates: for a fixed (x, y), the equation $Ax^2 + Bxy + Cy^2 + Dx + Ey + F = 0$ is a *linear* equation in (A, B, C, D, E, F).

- By dimension counting five constraints (that the curve passes through five points) are necessary to specify a conic, as each constraint cuts the dimension of possibilities by 1, and one starts with 5 dimensions.

- In 5 dimensions, the intersection of 5 (independent) hyperplanes is a single point (formally, by Bézout's theorem).

- General linear position of the points means that the constraints are *independent,* and thus do specify a unique conic.

- The resulting conic is non-degenerate because it is a curve (since it has more than 1 point), and does not contain a line (else it would split as two lines, at least one of which must contain 3 of the 5 points, by the pigeonhole principle), so it is irreducible.

The two subtleties in the above analysis are that the resulting point is a quadratic equation (not a linear equation), and that the constraints are independent. The first is simple: if A, B, and C all vanish, then the equation $Dx + Ey + F = 0$ defines a line, and any 3 points on this (indeed any number of points) lie on a line – thus general linear position ensures a conic. The second, that the constraints are independent, is significantly subtler: it corresponds to the fact that given five points in general linear position in the plane, their images in $\mathbf{P}^5$ under the Veronese map are in general linear position, which is true because the Veronese map is biregular: i.e., if the image of five points satisfy a relation, then the relation can be pulled back and the original points must also satisfy a relation. The Veronese map has coordinates $[x^2 : xy : y^2 : xz : yz : z^2]$, and the target $\mathbf{P}^5$ is *dual* to the $[A : B : C : D : E : F]$ $\mathbf{P}^5$ of conics. The Veronese map corresponds to "evaluation of a conic at a point", and the statement about independence of constraints is exactly a geometric statement about this map.

That five points determine a conic can be proven by synthetic geometry—i.e., in terms of lines and points in the plane—in addition to the analytic (algebraic) proof given above. Such a proof can be given using a theorem of Jakob Steiner, which states:

> "Given a projective transformation f, between the pencil of lines passing through a point X and the pencil of lines passing through a point Y, the set C of intersection points between a line x and its image $f(x)$ forms a conic".

> "Note that X and Y are on this conic by considering the preimage and image of the line XY (which is respectively a line through X and a line through Y)".

This can be shown by taking the points X and Y to the standard points $[1:0:0]$ and $[0:1:0]$ by a projective transformation, in which case the pencils of lines correspond to the horizontal and vertical lines in the plane, and the intersections of corresponding lines to the graph of a function, which (must be shown) is a hyperbola, hence a conic, hence the original curve C is a conic.

Now given five points X, Y, A, B, C, the three lines XA, XB, XC can be taken to the three lines YA, YB, YC by a unique projective transform, since projective transforms are simply 3-transitive on lines (they are simply 3-transitive on points, hence by projective duality they are 3-transitive on lines). Under this map X maps to Y, since these are the unique intersection points of these lines, and thus satisfy the hypothesis of Steiner's theorem. The resulting conic thus contains all five points, and is the unique such conic, as desired.

Construction

Analytically, given the coordinates of the five points, the equation for the conic can be found by linear algebra, by writing and solving the five equations in the coefficients, substituting the variables with the values of the coordinates: five equations, six unknowns, but homogeneous so scaling removes one dimension; concretely, setting one of the coefficients to 1 accomplishes this.

Synthetically, the conic can be constructed by the Braikenridge–Maclaurin construction, by applying the Braikenridge–Maclaurin theorem, which is the converse of Pascal's theorem. Pascal's theorem states that given 6 points on a conic (a hexagon), the lines defined by opposite sides intersect in three collinear points. This can be reversed to construct the possible locations for a 6th point, given 5 existing ones.

Generalizations

The natural generalization is to ask for what value of k a configuration of k points (in general position) in n-space determines a variety of degree d and dimension m, which is a fundamental question in enumerative geometry.

A simple case of this is for a hypersurface (a codimension 1 subvariety, the zeros of a single polynomial, the case $m = n-1$), of which plane curves are an example.

In the case of a hypersurface, the answer is given in terms of the multiset coefficient, more familiarly the binomial coefficient, or more elegantly the rising factorial, as:

$$k = \left(\left(\begin{array}{c} n+1 \\ d \end{array} \right) \right) - 1 = \left(\begin{array}{c} n+d \\ d \end{array} \right) - 1 = \frac{1}{n!}(d+1)^{(n)} - 1.$$

This is via the analogous analysis of the Veronese map: k points in general position impose k independent linear conditions on a variety (because the Veronese map is biregular), and the number of monomials of degree d in $n+1$ variables (n-dimensional projective space has $n+1$ homogeneous coordinates) is $\left(\begin{array}{c} n+1 \\ d \end{array} \right)$ from which 1 is subtracted because of projectivization: multiplying a polynomial by a constant does not change its zeros.

In the above formula, the number of points k is a polynomial in d of degree n, with leading coefficient $1/n$.

In the case of plane curves, where $n = 2$, the formula becomes:

$$\frac{1}{2}(d+1)(d+2) - 1 = \frac{1}{2}(d^2 + 3d)$$

whose values for $d = 0, 1, 2, 3, 4$ are $0, 2, 5, 9, 14$ – there are no curves of degree 0 (a single point is determines a point, which is codimension 2), 2 points determine a line, 5 points determine a conic, 9 points determine a cubic, 14 points determine a quartic, and so forth.

While five points determine a conic, sets of six or more points on a conic are not in general position, that is, they are constrained as is demonstrated in Pascal's theorem.

Similarly, while nine points determine a cubic, if the nine points lie on more than one cubic—i.e., they are the intersection of two cubics—then they are not in general position, and indeed satisfy an addition constraint, as stated in the Cayley–Bacharach theorem.

Four points do not determine a conic, but rather a pencil, the 1-dimensional linear system of conics which all pass through the four points (formally, have the four points as base locus). Similarly, three points determine a 2-dimensional linear system (net), two points determine a 3-dimensional linear system (web), one point determines a 4-dimensional linear system, and zero points place no constraints on the 5-dimensional linear system of all conics.

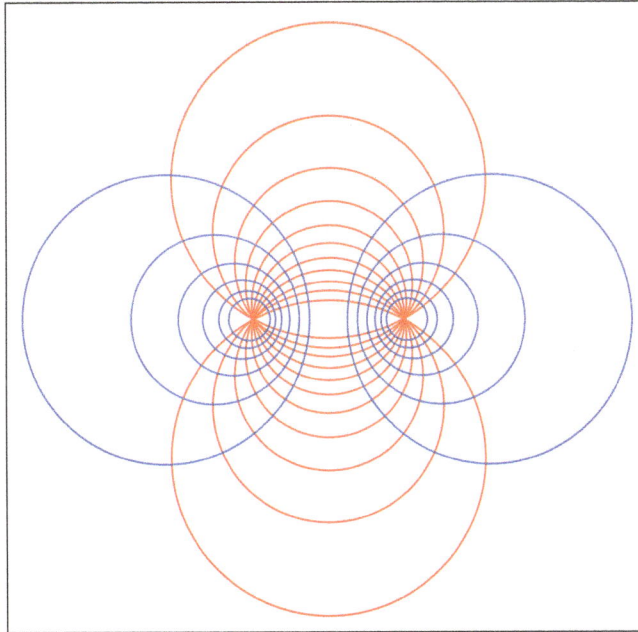

The Apollonian circles are two 1-parameter families determined by 2 points.

As is well known, three non-collinear points determine a circle in Euclidean geometry and two distinct points determine a pencil of circles such as the Apollonian circles. These results seem to run counter the general result since circles are special cases of conics. However, in a pappian projective plane a conic is a circle only if it passes through two specific points on the line at infinity, so a circle is determined by five non-collinear points, three in the affine plane and these two special points. Similar considerations explain the smaller than expected number of points needed to define pencils of circles.

Tangency

Instead of passing through points, a different condition on a curve is being tangent to a given line. Being tangent to five given lines also determines a conic, by projective duality, but from the algebraic point of view tangency to a line is a *quadratic* constraint, so naive dimension counting yields $2^5 = 32$ conics tangent to five given lines, of which 31 must be ascribed to degenerate conics, as described in fudge factors in enumerative geometry; formalizing this intuition requires significant further development to justify.

Another classic problem in enumerative geometry, of similar vintage to conics, is the Problem of Apollonius: a circle that is tangent to three circles in general determines eight circles, as each of these is a quadratic condition and $2^3 = 8$. As a question in real geometry, a full analysis involves many special cases, and the actual number of circles may be any number between 0 and 8, except for 7.

References

- Ptolemy-theorem, geometry, mathematics, personal-development: trans4mind.com, Retrieved 29 June, 2019

- Aigner, Martin; Ziegler, Günter M. (2014). Proofs from THE BOOK. Springer. Pp. 91–93. ISBN 9783540404606

- Brahmagupta, emt725: coe.uga.edu, Retrieved 13 March, 2019

- Alsina, Claudi; Nelsen, Roger B. (2010), "The Finsler–Hadwiger Theorem 8.5", Charming Proofs: A Journey Into Elegant Mathematics, Mathematical Association of America, p. 125, ISBN 9780883853481

5

Conic Section

Conic section is a shape which is formed by the intersection of a cone and a plane. A conic section can attain different shapes depending upon the angle of intersection such as of a circle, an ellipse, a parabola or a hyperbola. These diverse conic sections have been thoroughly discussed in this chapter.

Conic section, also called conic, in geometry, is any curve produced by the intersection of a plane and a right circular cone. Depending on the angle of the plane relative to the cone, the intersection is a circle, an ellipse, a hyperbola, or a parabola. Special (degenerate) cases of intersection occur when the plane passes through only the apex (producing a single point) or through the apex and another point on the cone (producing one straight line or two intersecting straight lines).

The basic descriptions, but not the names, of the conic sections can be traced to Menaechmus, a pupil of both Plato and Eudoxus of Cnidus. Apollonius of Perga, known as the "Great Geometer," gave the conic sections their names and was the first to define the two branches of the hyperbola (which presuppose the double cone). Apollonius's eight-volume treatise on the conic sections, Conics, is one of the greatest scientific works from the ancient world.

Conics may also be described as plane curves that are the paths (loci) of a point moving so that the ratio of its distance from a fixed point (the focus) to the distance from a fixed line (the directrix) is a constant, called the eccentricity of the curve. If the eccentricity is zero, the curve is a circle; if equal to one, a parabola; if less than one, an ellipse; and if greater than one, a hyperbola.

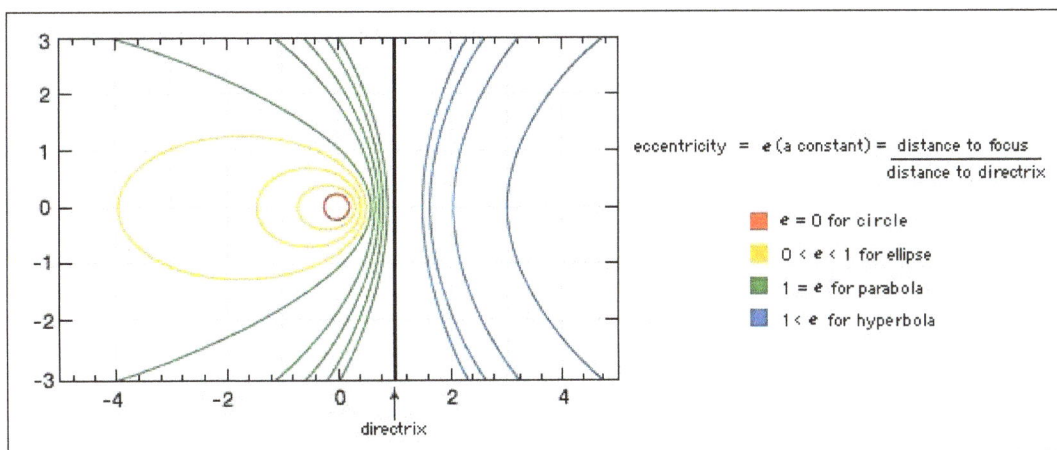

Eccentricity of conic sections. The eccentricity of a conic section completely characterizes its shape. For example, all circles have zero eccentricity, and all parabolas have unit eccentricity; hence, all circles (and all parabolas) have the same shape, only varying in size. (Under appropriate magnification they are indistinguishable.) In contrast, ellipses and hyperbolas vary greatly in shape.

Every conic section corresponds to the graph of a second degree polynomial equation of the form $Ax^2 + By^2 + 2Cxy + 2Dx + 2Ey + F = 0$, where x and y are variables and A, B, C, D, E, and F are coefficients that depend upon the particular conic. By a suitable choice of coordinate axes, the equation for any conic can be reduced to one of three simple r forms:

$$\frac{x^2}{a^2} + \frac{y^2}{b^2} = 1, \quad \frac{x^2}{a^2} - \frac{y^2}{b^2} = 1, \quad \text{or } y^2 = 2px,$$

corresponding to an ellipse, a hyperbola, and a parabola, respectively. (An ellipse where a = b is in fact a circle.) The extensive use of coordinate systems for the algebraic analysis of geometric curves originated with René Descartes.

Post-greek Applications

Conic sections found their first practical application outside of optics in 1609 when Johannes Kepler derived his first law of planetary motion: A planet travels in an ellipse with the Sun at one focus. Galileo Galilei published the first correct description of the path of projectiles—a parabola—in his Dialogues of the Two New Sciences. In 1639 the French engineer Girard Desargues initiated the study of those properties of conics that are invariant under projections. Eighteenth-century architects created a fad for whispering galleries—such as in the U.S. Capital and in St. Paul's Cathedral in London—in which a whisper at one focus of an ellipsoid (an ellipse rotated about one axis) can be heard at the other focus, but nowhere else. From the ubiquitous parabolic satellite dish to the use of ultrasound in lithotripsy, new applications for conic sections continue to be found.

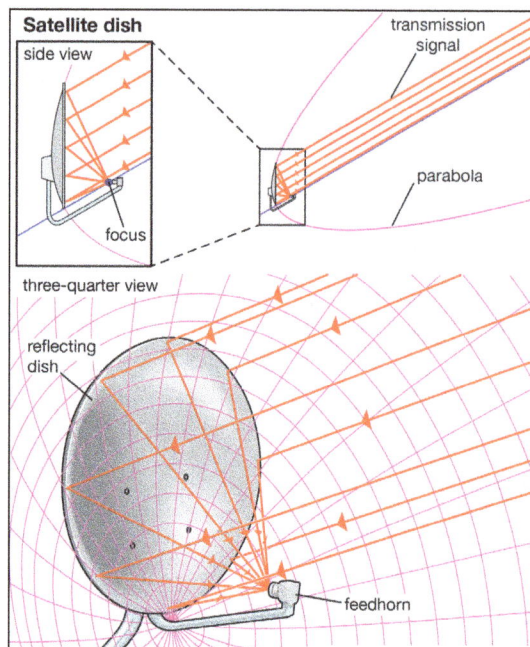

Satellite dish
side view
transmission signal
focus
parabola
three-quarter view
reflecting dish
feedhorn

Parabolic satellite dish antennaSatellite dishes are often shaped like portions of a paraboloid (a parabola rotated about its central axis) in order to focus transmission signals onto the pickup receiver, or feedhorn. Typically, the section of the paraboloid used is offset from the centre so that the feedhorn and its support do not unduly block signals to the reflecting dish.

HYPERBOLA

In mathematics, a hyperbola is a type of smooth curve lying in a plane, defined by its geometric properties or by equations for which it is the solution set. A hyperbola has two pieces, called connected components or branches, that are mirror images of each other and resemble two infinite bows. The hyperbola is one of the three kinds of conic section, formed by the intersection of a plane and a double cone. (The other conic sections are the parabola and the ellipse. A circle is a special case of an ellipse.) If the plane intersects both halves of the double cone but does not pass through the apex of the cones, then the conic is a hyperbola.

Hyperbolas arise in many ways:

- As the curve representing the function $f(x) = 1/x$ in the Cartesian plane.

- As the path followed by the shadow of the tip of a sundial.

- As the shape of an open orbit (as distinct from a closed elliptical orbit), such as the orbit of a spacecraft during a gravity assisted swing-by of a planet or more generally any spacecraft exceeding the escape velocity of the nearest planet.

- As the path of a single-apparition comet (one travelling too fast ever to return to the solar system).

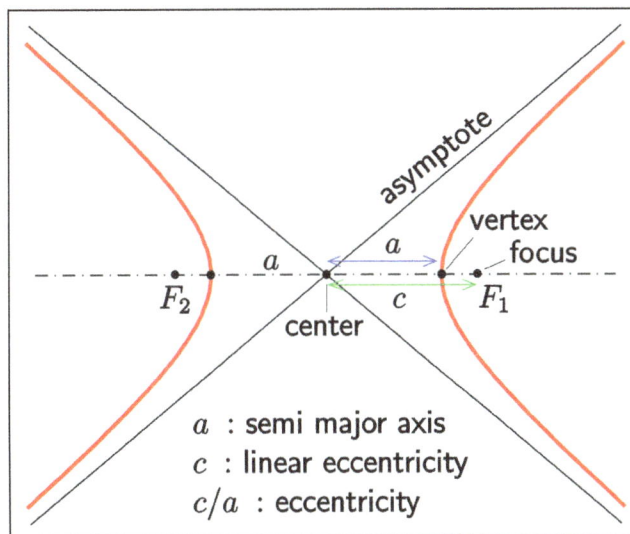

Hyperbola (red): features.

- As the scattering trajectory of a subatomic particle (acted on by repulsive instead of attractive forces but the principle is the same).

- In radio navigation, when the difference between distances to two points, but not the distances themselves, can be determined, and so on.

Each branch of the hyperbola has two arms which become straighter (lower curvature) further out from the center of the hyperbola. Diagonally opposite arms, one from each branch, tend in the limit to a common line, called the asymptote of those two arms. So there are two asymptotes, whose intersection is at the center of symmetry of the hyperbola, which can be thought of as the mirror point about which each branch reflects to form the other branch. In the case of the curve $f(x) = 1/x$ the asymptotes are the two coordinate axes.

Hyperbolas share many of the ellipses' analytical properties such as eccentricity, focus, and directrix. Typically the correspondence can be made with nothing more than a change of sign in some term. Many other mathematical objects have their origin in the hyperbola, such as hyperbolic paraboloids (saddle surfaces), hyperboloids ("wastebaskets"), hyperbolic geometry (Lobachevsky's celebrated non-Euclidean geometry), hyperbolic functions (sinh, cosh, tanh, etc.), and gyrovector spaces (a geometry proposed for use in both relativity and quantum mechanics which is not Euclidean).

Hyperbola as Locus of Points

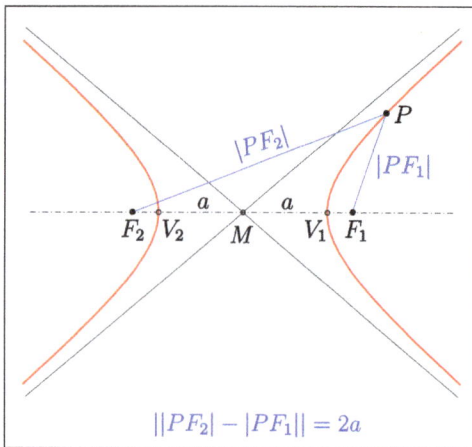

Hyperbola: definition by the distances of points to two fixed points (foci).

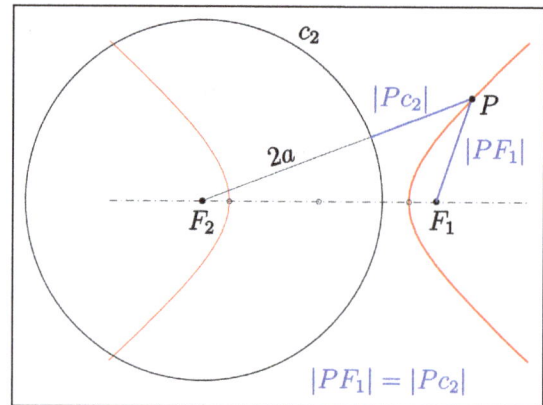

Hyperbola: definition with circular directrix.

A hyperbola can be defined geometrically as a set of points (locus of points) in the Euclidean plane:

A hyperbola is a set of points, such that for any point p of the set, the absolute difference of the distances $|PF_1|, |PF_2|$ to two fixed points F_1, F_2 (the *foci*), is constant, usually denoted by $2a, a > 0$:

$$H = \{P \mid \|PF_2\| - |PF_1\| = 2a\}.$$

The midpoint M of the line segment joining the foci is called the *center* of the hyperbola. The line through the foci is called the *major axis*. It contains the *vertices* V_1, V_2, which have distance a to the center. The distance c of the foci to the center is called the *focal distance* or *linear eccentricity*. The quotient $\frac{c}{a}$ is the *eccentricity e*.

The equation $\| PF_2 | - | PF_1 \| = 2a$ can be viewed in a different way:

If c_2 is the circle with midpoint F_2 and radius $2a$, then the distance of a point p of the right branch to the circle c_2 equals the distance to the focus F_1:

$$| PF_1 | = | Pc_2 |.$$

c_2 is called the *circular directrix* (related to focus F_2) of the hyperbola. In order to get the left branch of the hyperbola, one has to use the circular directrix related to F_1.

Hyperbola in Cartesian Coordinates

If Cartesian coordinates are introduced such that the origin is the center of the hyperbola and the x-axis is the major axis, then the hyperbola is called *east-west-opening* and the *foci* are the points $F_1 = (c,0), F_2 = (-c,0)$, the *vertices* are $V_1 = (a,0), V_2 = (-a,0)$.

For an arbitrary point (x, y) the distance to the focus $(c,0)$ is $\sqrt{(x-c)^2 + y^2}$ and to the second focus $\sqrt{(x+c)^2 + y^2}$. Hence the point (x, y) is on the hyperbola if the following condition is fulfilled:

$$\sqrt{(x-c)^2 + y^2} - \sqrt{(x+c)^2 + y^2} = \pm 2a.$$

Remove the square roots by suitable squarings and use the relation $b^2 = c^2 - a^2$ to obtain the equation of the hyperbola:

$$\frac{x^2}{a^2} - \frac{y^2}{b^2} = 1.$$

This equation is called the *canonical form* of a hyperbola, because any hyperbola, regardless of its orientation relative to the Cartesian axes and regardless of the location of its center, can be transformed to this form by a change of variables, giving a hyperbola that is congruent to the original.

The axes of symmetry or *principal axes* are the *transverse axis* (containing the segment of length $2a$ with endpoints at the vertices) and the *conjugate axis* (containing the segment of length $2b$ perpendicular to the transverse axis and with midpoint at the hyperbola's center). As opposed to an ellipse, a hyperbola has only two vertices: $(a,0), (-a,0)$. The two points $(0,b), (0,-b)$ on the conjugate axis are *not* on the hyperbola.

It follows from the equation that the hyperbola is *symmetric* with respect to both of the coordinate axes and hence symmetric with respect to the origin.

Eccentricity

For a hyperbola in the above canonical form, the eccentricity is given by,

$$e = \sqrt{1 + \frac{b^2}{a^2}}.$$

Two hyperbolas are geometrically similar to each other – meaning that they have the same shape, so that one can be transformed into the other by rigid left and right movements, rotation, taking a mirror image, and scaling (magnification) – if and only if they have the same eccentricity.

Asymptotes

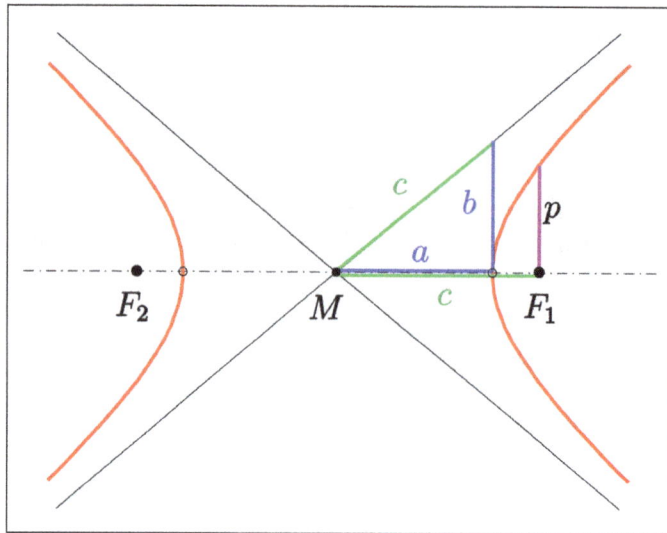

Hyperbola: semi-axes a,b, linear eccentricity c, semi latus rectum p.

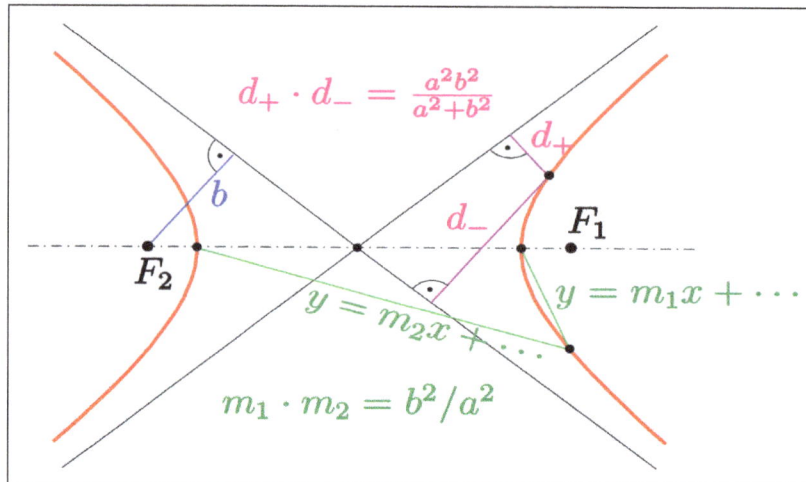

Hyperbola: 3 properties.

Solving the equation (above) of the hyperbola for y yields:

$$y = \pm \frac{b}{a}\sqrt{x^2 - a^2}.$$

It follows from this that the hyperbola approaches the two lines:

$$y = \pm \frac{b}{a}x$$

for large values of $|x|$. These two lines intersect at the center (origin) and are called asymptotes:

of the hyperbola $\frac{x^2}{a^2} - \frac{y^2}{b^2} = 1$.

With help of the figure one can see that:

The distance from a focus to either asymptote is b (the semi-minor axis). From the Hesse normal form $\frac{bx \pm ay}{\sqrt{a^2 + b^2}} = 0$ of the asymptotes and the equation of the hyperbola one gets.

The product of the distances from a point on the hyperbola to both the asymptotes is the constant $\frac{a^2 b^2}{a^2 + b^2}$, which can also be written in terms of the eccentricity e as $\left(\frac{b}{e}\right)^2$.

From the equation $y = \pm \frac{b}{a}\sqrt{x^2 - a^2}$ of the hyperbola (above) one can derive.

The product of the slopes of lines from a point P to the two vertices is the constant b^2 / a^2.

The product of the distances from a point on the hyperbola to the asymptotes along lines parallel to the asymptotes is the constant $\frac{a^2 + b^2}{4}$.

Semi-latus Rectum

The length of the chord through one of the foci, perpendicular to the major axis of the hyperbola, is called the *latus rectum*. One half of it is the *semi-latus rectum b*. A calculation shows:

$$p = \frac{b^2}{a}.$$

The semi-latus rectum p may also be viewed as the *radius of curvature* of the osculating circles at the vertices.

Tangent

The simplest way to determine the equation of the tangent at a point (x_0, y_0) is to implicitly differentiate the equation $\frac{x^2}{a^2} - \frac{y^2}{b^2} = 1$ of the hyperbola. Denoting dy/dx as y', this produces:

$$\frac{2x}{a^2} - \frac{2yy'}{b^2} = 0 \Rightarrow y' = \frac{x}{y}\frac{b^2}{a^2} \Rightarrow y = \frac{x_0}{y_0}\frac{b^2}{a^2}(x - x_0) + y_0.$$

With respect to $\frac{x_0^2}{a^2} - \frac{y_0^2}{b^2} = 1$, the equation of the tangent at point (x_0, y_0) is:

$$\frac{x_0}{a^2}x - \frac{y_0}{b^2}y = 1.$$

A particular tangent line distinguishes the hyperbola from the other conic sections. Let f be the distance from the vertex V (on both the hyperbola and its axis through the two foci) to the nearer focus. Then the distance, along a line perpendicular to that axis, from that focus to a point P on the hyperbola is greater than $2f$. The tangent to the hyperbola at P intersects that axis at point Q at an angle $\angle PQV$ of greater than $45°$.

Rectangular Hyperbola

In the case $a = b$ the hyperbola is called *rectangular* (or *equilateral*), because its asymptotes intersect rectangularly (that is, are perpendicular). For this case, the linear eccentricity is $c = \sqrt{2}ac$, the eccentricity $e = \sqrt{2}$ and the semi-latus rectum $p = a$.

Parametric Representation with Hyperbolic Sine/Cosine

Using the hyperbolic sine and cosine functions $\cosh, \sinh$, a parametric representation of the hyperbola $\frac{x^2}{a^2} - \frac{y^2}{b^2} = 1$ can be obtained, which is similar to the parametric representation of an ellipse:

$$\left(\pm a \cosh t , b \sinh t \right), t \in \mathbb{R}$$

which satisfies the Cartesian equation because $\cosh^2 t - \sinh^2 t = 1$.

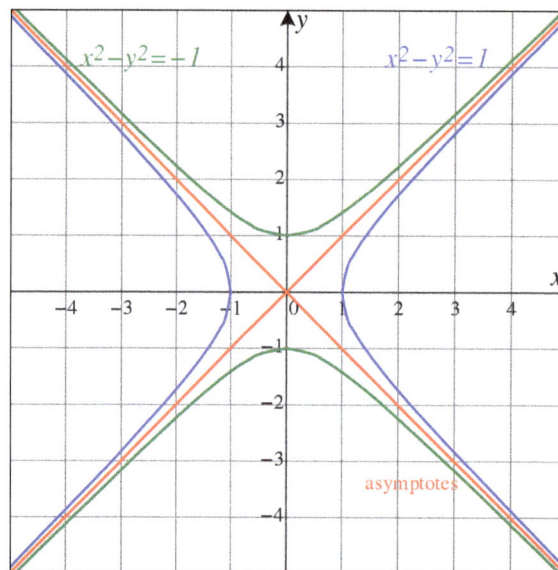

Here $a = b = 1$ giving the unit hyperbola in blue and its conjugate hyperbola in green, sharing the same red asymptotes.

Conjugate Hyperbola

Exchange x and y to obtain the equation of the conjugate hyperbola:

$$\frac{y^2}{a^2} - \frac{x^2}{b^2} = 1 , \text{ also written as:}$$

$$\frac{x^2}{b^2} - \frac{y^2}{a^2} = -1 .$$

Hyperbolic Functions

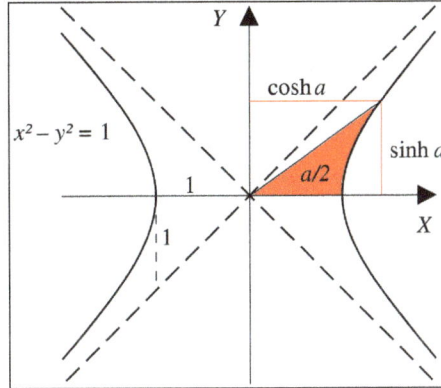

A ray through the unit hyperbola $x^2 - y^2 = 1$ in the point $(\cosh a, \sinh a)$, where a is twice the area between the ray, the hyperbola, and the x-axis. For points on the hyperbola below the $-x$ axis, the area is considered negative.

Just as the trigonometric functions are defined in terms of the unit circle, so also the hyperbolic functions are defined in terms of the unit hyperbola, as shown in this diagram.

Let a be twice the area between the x axis and a ray through the origin intersecting the unit hyperbola and define $b = \cosh(a)$ as the horizontal coordinate of the intersection point. Then, by the Pythagorean theorem,

$$\frac{a}{2} = \frac{b\sqrt{b^2-1}}{2} - \int_1^b \sqrt{x^2-1}\,dx = \frac{b\sqrt{b^2-1}}{2} - \frac{b\sqrt{b^2-1} - \ln\left(b+\sqrt{b^2-1}\right)}{2},$$

which simplifies to:

$$a = \ln\left(b+\sqrt{b^2-1}\right).$$

Solving for b yields the exponential form of the hyperbolic cosine:

$$x = b = \cosh a = \frac{e^a + e^{-a}}{2}.$$

By defining y as $\sinh a$, from $x^2 - y^2 = 1$ one gets:

$$\sinh a = \sqrt{\cosh^2 a - 1}.$$

This implies the exponential form of the hyperbolic sine (h is the vertical coordinate of the intersection point):

$$y = h = \sinh a = \frac{e^a - e^{-a}}{2}.$$

Other hyperbolic functions are defined according to the hyperbolic cosine and hyperbolic sine, so for example,

$$\tanh a = \frac{\sinh a}{\cosh a} = \frac{e^{2a}-1}{e^{2a}+1}.$$

The area hyperbolic cosine is defined as follows:

$$a = \operatorname{arcosh} b = \ln\left(b + \sqrt{b^2 - 1}\right).$$

Solving for a in the hyperbolic sine equation yields the area hyperbolic sine:

$$a = \operatorname{arsinh} h = \ln\left(h + \sqrt{h^2 + 1}\right).$$

Hyperbola with equation $y = A/x$.

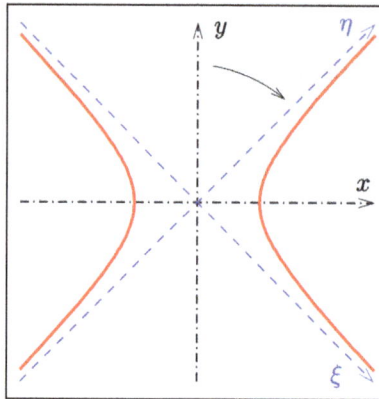

Rotating the coordinate system in order to describe a rectangular hyperbola as graph of a function.

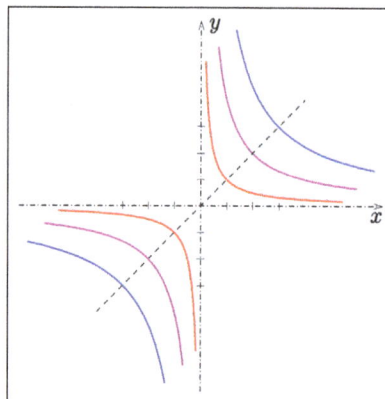

Three rectangular hyperbolas $y = A / x$ with the coordinate axes
as asymptotes red: $A = 1$; magenta: $A = 4$; blue: $A = 9$.

If the xy-coordinate system is rotated about the origin by the angle $-45°$ and new coordinates ξ, η are assigned, then $x = \frac{\xi+\eta}{\sqrt{2}}$, $y = \frac{-\xi+\eta}{\sqrt{2}}$. The rectangular hyperbola $\frac{x^2-y^2}{a^2} = 1$ (whose semi-axes are equal) has the new equation $\frac{2\xi\eta}{a^2} = 1$. Solving for η yields $\eta = \frac{a^2/2}{\xi}$.

Thus, in an xy-coordinate system the graph of a function $f : x \mapsto \frac{A}{x}$, $A > 0$, with equation,

$y = \dfrac{A}{x}$, $A > 0$, is a *rectangular hyperbola* entirely in the first and third quadrants with:

- The coordinate axes as asymptotes.

- The line $y = x$ as major axis.

- The center $(0,0)$ and the semi-axis $a = b = \sqrt{2A}$.

- The vertices $\left(\sqrt{A}, \sqrt{A}\right), \left(-\sqrt{A}, -\sqrt{A}\right)$.

- The semi-latus rectum and radius of curvature at the vertices $p = a = \sqrt{2A}$.

- The linear eccentricity $c = 2\sqrt{A}$ and the eccentricity $e = \sqrt{2}$.

- The tangent $y = -\dfrac{A}{x_0^2} x + 2\dfrac{A}{x_0}$ at point $(x_0, A/x_0)$.

A rotation of the original hyperbola by $+45°$ results in a rectangular hyperbola entirely in the second and fourth quadrants, with the same asymptotes, center, semi-latus rectum, radius of curvature at the vertices, linear eccentricity, and eccentricity as for the case of $-45°$ rotation, with equation:

$$y = \dfrac{-A}{x}, A > 0,$$

- The *semi-axes* $a = b = \sqrt{2A}$,

- The line $y = -x$ as major axis,

- The *vertices* $\left(-\sqrt{A}, \sqrt{A}\right), \left(\sqrt{A}, -\sqrt{A}\right)$.

Shifting the hyperbola with equation $y = \dfrac{A}{x}$, $A \neq 0$, so that the new center is (c_0, d_0), yields the new equation:

$$y = \dfrac{A}{x - c_0} + d_0,$$

and the new asymptotes are $x = c_0$ and $y = d_0$. The shape parameters a, b, p, c, e remain unchanged.

Hyperbola by the Directrix Property

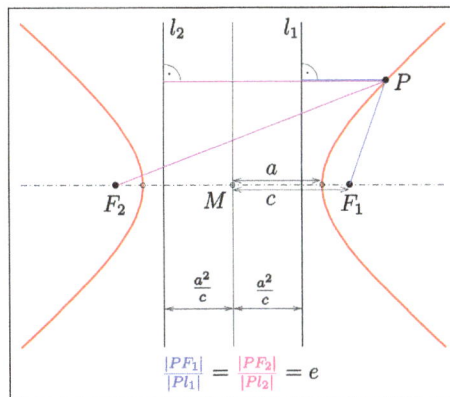

Hyperbola: directrix property.

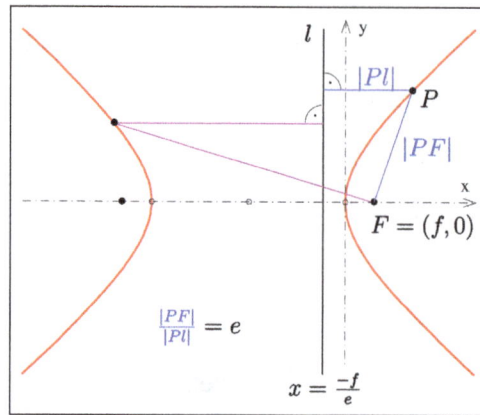

Hyperbola: definition with directrix property.

The two lines at distance $d = \dfrac{a^2}{c}$ and parallel to the minor axis are called directrices of the hyperbola.

For an arbitrary point P of the hyperbola the quotient of the distance to one focus and to the corresponding directrix is equal to the eccentricity:

$$\frac{|PF_1|}{|Pl_1|} = \frac{|PF_2|}{|Pl_2|} = e = \frac{c}{a}.$$

The proof for the pair F_1, l_1 follows from the fact that $|PF_1|^2 = (x-c)^2 + y^2, |Pl_1|^2 = \left(x - \frac{a^2}{c}\right)^2$ and $y^2 = \frac{b^2}{a^2}x^2 - b^2$ satisfy the equation:

$$|PF_1|^2 - \frac{c^2}{a^2}|Pl_1|^2 = 0$$

The second case is proven analogously.

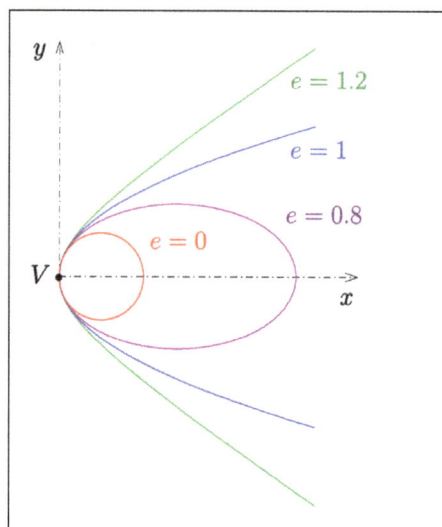

Pencil of conics with a common vertex and common semi latus rectum.

The inverse statement is also true and can be used to define a hyperbola.

For any point F (focus), any line l (directrix) not through F and any real number e with $e > 1$ the set of points (locus of points), for which the quotient of the distances to the point and to the line is e is a hyperbola.

$$H = \left\{ P \Big| \frac{|PF|}{|Pl|} = e \right\}$$

(The choice $e = 1$ yields a parabola and if $e < 1$ an ellipse).

Let $F = (f, 0)$, $e > 0$ and assume $(0,0)$ is a point on the curve. The directrix l has equation $x = -\frac{f}{e}$. With $P = (x, y)$, the relation $|PF|^2 = e^2 |Pl|^2$ produces the equations:

$$(x - f)^2 + y^2 = e^2 \left(x + \frac{f}{e} \right)^2 = (ex + f)^2 \text{ and } x^2(e^2 - 1) + 2xf(1 + e) - y^2 = 0.$$

The substitution $p = f(1 + e)$ yields:

$$x^2(e^2 - 1) + 2px - y^2 = 0.$$

This is the equation of an *ellipse* ($e < 1$) or a *parabola* ($e = 1$) or a *hyperbola* $e > 1$. All of these non-degenerate conics have, in common, the origin as a vertex.

If $e > 1$, introduce new parameters a, b so that $e^2 - 1 = \frac{b^2}{a^2}$, and $p = \frac{b^2}{a}$, and then the equation above becomes:

$$\frac{(x + a)^2}{a^2} - \frac{y^2}{b^2} = 1 ,$$

which is the equation of a hyperbola with center $(-a, 0)$, the x-axis as major axis and the major/minor semi axis a, b.

Hyperbola as Plane Section of a Cone

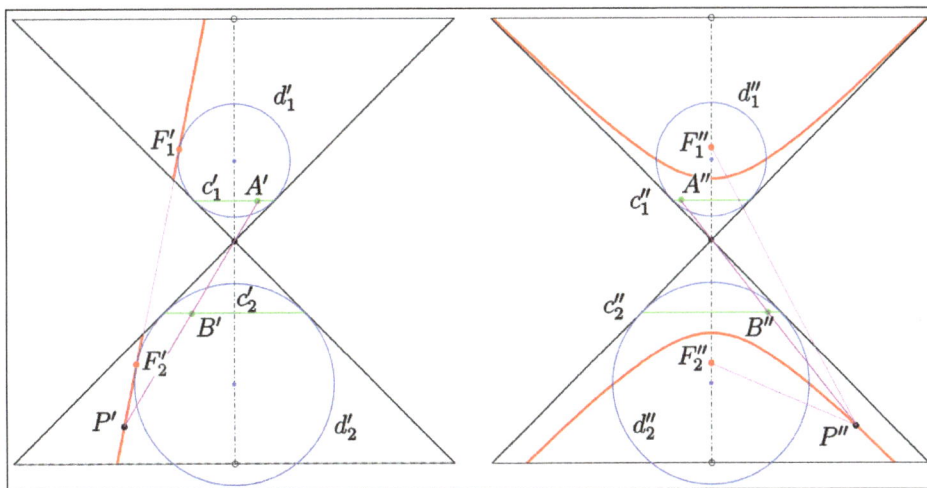

Hyperbola (red): two views of a cone and two Dandelin spheres d_1, d_2.

The intersection of an upright double cone by a plane not through the vertex with slope greater than the slope of the lines on the cone is a hyperbola. In order to prove the defining property of a hyperbola one uses two Dandelin spheres d_1, d_2, which are spheres that touch the cone along circles c_1, c_2 and the intersecting (hyperbola) plane at points F_1 and F_2. It turns out: F_1, F_2 are the *foci* of the hyperbola.

- Let P be an arbitrary point of the intersection curve.

- The generator (line) of the cone containing P intersects circle c_1 at point A and circle c_2 at a point B.

- The line segments PF_1 and PA are tangential to the sphere d_1 and, hence, are of equal length.

- The line segments PF_2 and $\overline{PB}$ are tangential to the sphere d_2 and, hence, are of equal length.

- The result is: $|PF_1| - |PF_2| = |PA| - |PB| = |AB|$ is independent of the hyperbola point P.

Pin and String Construction

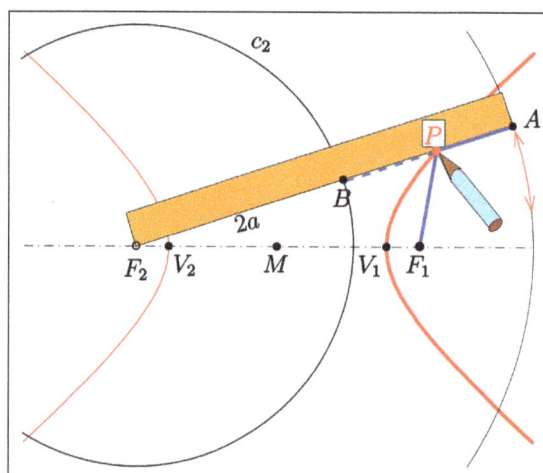

Hyperbola: Pin and string construction.

The definition of a hyperbola by its foci and its circular directrices can be used for drawing an arc of it with help of pins, a string and a ruler:

- Choose the *foci* F_1, F_2, the vertices V_1, V_2 and one of the *circular directrices*, for example c_2 (circle with radius $2a$).

- A *ruler* is fixed at point F_2 free to rotate around F_2. Point B is marked at distance $2a$.

- A *string* with length $|AB|$ is prepared.

- One end of the string is pinned at point A on the ruler, the other end is pinned to point F_1.

- Take a *pen* and hold the string tight to the edge of the ruler.

- *Rotating* the ruler around F_2 prompts the pen to draw an arc of the right branch of the hyperbola, because of $|PF_1| = |PB|$.

The Tangent Bisects the Angle between the Lines to the Foci

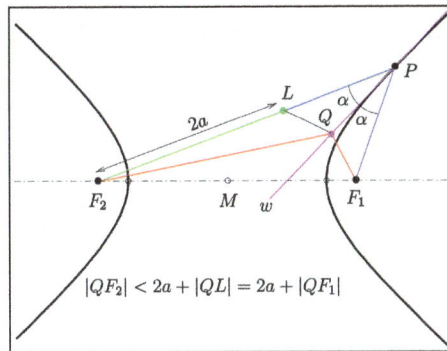

Hyperbola: the tangent bisects the lines through the foci.

The tangent at a point P bisects the angle between the lines $\overline{PF_1}, \overline{PF_2}$.

Let L be the point on the line $\overline{PF_2}$ with the distance $2a$ to the focus F_2. Line ω is the bisector of the angle between the lines $\overline{PF_1}, \overline{PF_2}$. In order to prove that ω is the tangent line at point P, one checks that any point Q on line ω which is different from P cannot be on the hyperbola. Hence ω has only point P in common with the hyperbola and is, therefore, the tangent at point P. From the diagram and the triangle inequality one recognizes that $|QF_2| < |LF_2| + |QL| = 2a + |QF_1|$ holds, which means: $|QF_2| - |QF_1| < 2a$. But if is a point of the hyperbola, the difference should be $2a$.

Midpoints of Parallel Chords

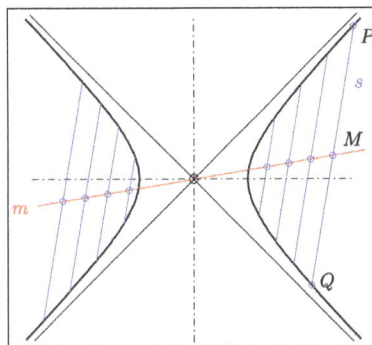

Hyperbola: the midpoints of parallel chords lie on a line.

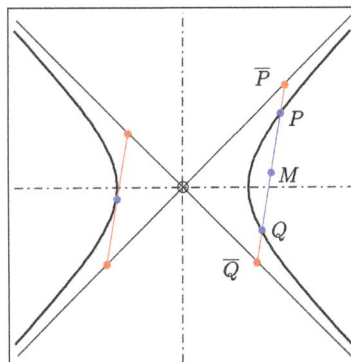

Hyperbola: the midpoint of a chord is the midpoint of the corresponding chord of the asymptotes.

- The midpoints of parallel chords of a hyperbola lie on a line through the center.

- The points of any chord may lie on different branches of the hyperbola.

The proof of the property on midpoints is best done for the hyperbola $y = 1/x$. Because any hyperbola is an affine image of the hyperbola $y = 1/x$ and an affine transformation preserves parallelism and midpoints of line segments, the property is true for all hyperbolas: For two points $P = \left(x_1, \frac{1}{x_1}\right), Q = \left(x_2, \frac{1}{x_2}\right)$ of the hyperbola $y = 1/x$:

- The midpoint of the chord is $M = \left(\frac{x_1 + x_2}{2}, \cdots\right) = \cdots = \frac{x_1 + x_2}{2}\left(1, \frac{1}{x_1 x_2}\right)$;

- The slope of the chord is $\dfrac{\frac{1}{x_2} - \frac{1}{x_1}}{x_2 - x_1} = \cdots = -\dfrac{1}{x_1 x_2}$.

For parallel chords the slope is constant and the midpoints of the parallel chords lie on the line $y = \frac{1}{x_1 x_2} x$.

Consequence: for any pair of points P, Q of a chord there exists a *skew reflection* with an axis (set of fixed points) passing through the center of the hyperbola, which exchanges the points P, Q and leaves the hyperbola (as a whole) fixed. A skew reflection is a generalization of an ordinary reflection across a line m, where all point-image pairs are on a line perpendicular to m.

Because a skew reflection leaves the hyperbola fixed, the pair of asymptotes is fixed, too. Hence the midpoint M of a chord P, Q divides the related line segment $\overline{PQ}$ between the asymptotes into halves, too. This means that $|P\overline{P}| = |Q\overline{Q}|$. This property can be used for the construction of further points Q of the hyperbola if a point P and the asymptotes are given.

If the chord degenerates into a *tangent*, then the touching point divides the line segment between the asymptotes in two halves.

Steiner Generation of a Hyperbola

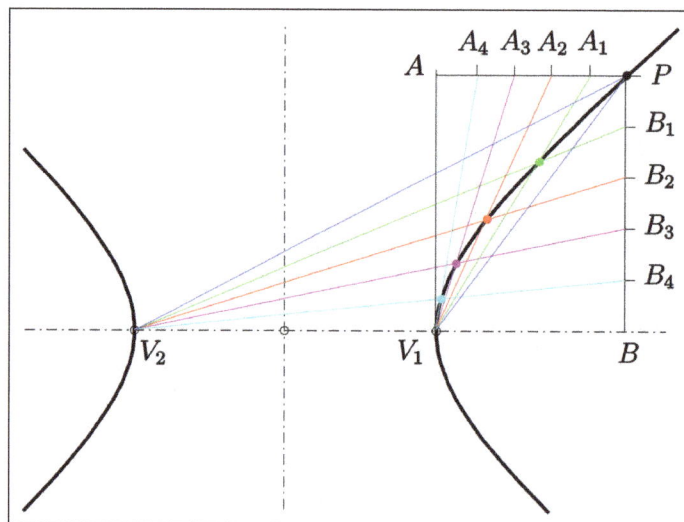

Hyperbola: Steiner generation.

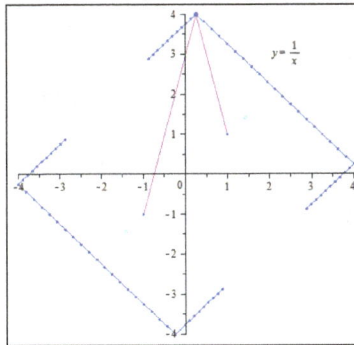

Hyperbola $y = 1/x$: Steiner generation.

The following method to construct single points of a hyperbola relies on the Steiner generation of a non degenerate conic section:

> "Given two pencils $B(U), B(V)$ of lines at two points U, V (all lines containing U and V, respectively) and a projective but not perspective mapping π of $B(U)$ onto $B(V)$, then the intersection points of corresponding lines form a non-degenerate projective conic section".

For the generation of points of the hyperbola $\frac{x^2}{a^2} - \frac{y^2}{b^2} = 1$ one uses the pencils at the vertices V_1, V_2. Let $P = (x_0, y_0)$ be a point of the hyperbola and $A = (a, y_0), B = (x_0, 0)$. The line segment $\overline{BP}$ is divided into n equally-spaced segments and this division is projected parallel with the diagonal AB as direction onto the line segment $\overline{AP}$. The parallel projection is part of the projective mapping between the pencils at V_1 and V_2 needed. The intersection points of any two related lines $S_1 A_i$ and $S_2 B_i$ are points of the uniquely defined hyperbola.

The subdivision could be extended beyond the points A and B in order to get more points, but the determination of the intersection points would become more inaccurate. A better idea is extending the points already constructed by symmetry.

1. The Steiner generation exists for ellipses and parabolas, too.

2. The Steiner generation is sometimes called a *parallelogram method* because one can use other points rather than the vertices, which starts with a parallelogram instead of a rectangle.

Inscribed angles for hyperbolas y = a/(x – b) + c and the 3-point-form:

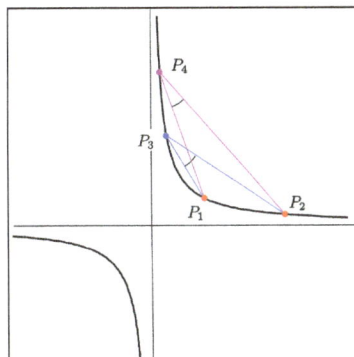

Hyperbola: inscribed angle theorem.

A hyperbola with equation $y = \frac{a}{x-b} + c, a \neq 0$ is uniquely determined by three points $(x_1, y_1), (x_2, y_2), (x_3, y_3)$ with different x- and y-coordinates. A simple way to determine the shape parameters a, b, c uses the *inscribed angle theorem* for hyperbolas:

In order to measure an angle between two lines with equations:

$$y = m_1 x + d_1, \ y = m_2 x + d_2 \ , m_1, m_2 \neq 0,$$

in this context one uses the quotient $\dfrac{m_1}{m_2}$.

Inscribed Angle Theorem for Hyperbolas

For four points $P_i = (x_i, y_i), i = 1, 2, 3, 4, \ x_i \neq x_k, y_i \neq y_k, i \neq k$ the following statement is true:

The four points are on a hyperbola with equation $y = \frac{a}{x-b} + c$ if and only if the angles at P_3 and P_4 are equal in the sense of the measurement above. That means if:

$$\frac{(y_4 - y_1)}{(x_4 - x_1)} \frac{(x_4 - x_2)}{(y_4 - y_2)} = \frac{(y_3 - y_1)}{(x_3 - x_1)} \frac{(x_3 - x_2)}{(y_3 - y_2)}$$

3-point-form of a Hyperbola's Equation

The equation of the hyperbola determined by 3 points $P_i = (x_i, y_i), i = 1, 2, 3, \ x_i \neq x_k, y_i \neq y_k, i \neq k$ is the solution of the equation for   .

$$\frac{(y - y_1)}{(x - x_1)} \frac{(x - x_2)}{(y - y_2)} = \frac{(y_3 - y_1)}{(x_3 - x_1)} \frac{(x_3 - x_2)}{(y_3 - y_2)}$$

Orthogonal Tangents – Orthoptic

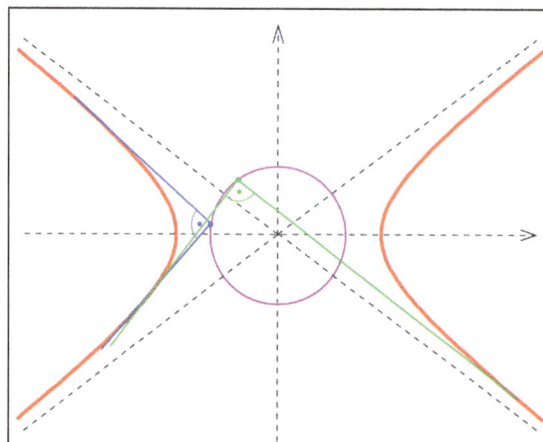

Hyperbola with its orthoptic (magenta).

For a hyperbola $\dfrac{x^2}{a^2} - \dfrac{y^2}{b^2} = 1, a > b$ the intersection points of *orthogonal* tangents lie on the circle $x^2 + y^2 = a^2 - b^2$. This circle is called the *orthoptic* of the given hyperbola. The tangents may belong to points on different branches of the hyperbola. In case of $a \leq b$ there are no pairs of orthogonal tangents.

Pole-polar Relation for a Hyperbola

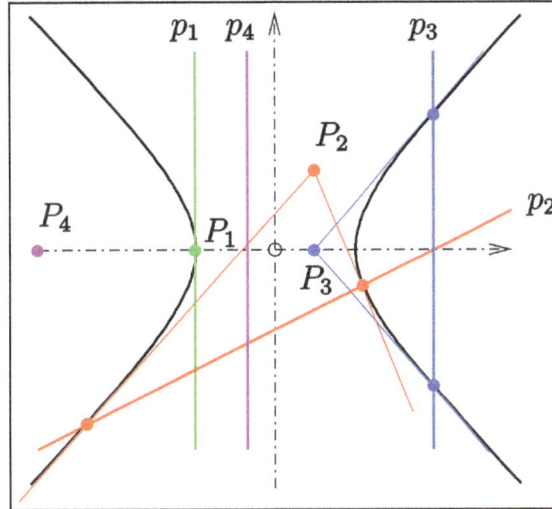

Hyperbola: pole-polar relation.

Any hyperbola can be described in a suitable coordinate system by an equation $\frac{x^2}{a^2} - \frac{y^2}{b^2} = 1$. The equation of the tangent at a point $P_0 = (x_0, y_0)$ of the hyperbola is $\frac{x_0 x}{a^2} - \frac{y_0 y}{b^2} = 1$. If one allows point $P_0 = (x_0, y_0)$ to be an arbitrary point different from the origin, then point $P_0 = (x_0, y_0) \neq (0,0)$ is mapped onto the line $\frac{x_0 x}{a^2} - \frac{y_0 y}{b^2} = 1$, not through the center of the hyperbola.

This relation between points and lines is a bijection.

The inverse function maps:

line $y = mx + d$, $d \neq 0$ onto the point $\left(-\frac{ma^2}{d}, -\frac{b^2}{d} \right)$ and

line $x = c$, $c \neq 0$ onto the point $\left(\frac{a^2}{c}, 0 \right)$.

Such a relation between points and lines generated by a conic is called pole-polar relation or just *polarity*. The pole is the point, the polar the line. By calculation one checks the following properties of the pole-polar relation of the hyperbola:

- For a point (pole) *on* the hyperbola the polar is the tangent at this point.

- For a pole P *outside* the hyperbola the intersection points of its polar with the hyperbola are the tangency points of the two tangents passing P.

- For a point *within* the hyperbola the polar has no point with the hyperbola in common.

- The intersection point of two polars (for example: p_2, p_3) is the pole of the line through their poles (here: P_2, P_3).

- The foci $(c,0)$, and $(-c,0)$ respectively and the directrices $x=\frac{a^2}{c}$ and $x=-\frac{a^2}{c}$ respectively belong to pairs of pole and polar.

Pole-polar relations exist for ellipses and parabolas, too.

Hyperbola as an Affine Image of the Unit Hyperbola x² − y² = 1

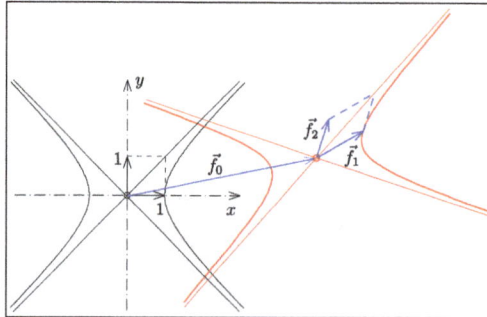

Hyperbola as an affine image of the unit hyperbola.

Any *hyperbola* is the affine image of the unit hyperbola with equation $x^2-y^2=1$.

An affine transformation of the Euclidean plane has the form $\vec{x}\to\vec{f_0}+A\vec{x}$, where A is a regular matrix (its determinant is not 0) and $\vec{f_0}$ is an arbitrary vector. If $\vec{f_1},\vec{f_2}$ are the column vectors of the matrix A, the unit hyperbola $(\pm\cosh(t),\sinh(t)),t\in\mathbb{R}$ is mapped onto the hyperbola:

$$\vec{x}=\vec{p}(t)=\vec{f_0}\pm\vec{f_1}\cosh t+\vec{f_2}\sinh t.$$

$\vec{f_0}$ is the center, $\vec{f_0}+\vec{f_1}$ a point of the hyperbola and $\vec{f_2}$ a tangent vector at this point. In general the vectors $\vec{f_1},\vec{f_2}$ are not perpendicular. That means, in general $\vec{f_0}\pm\vec{f_1}$ are *not* the vertices of the hyperbola. But $\vec{f_1}\pm\vec{f_2}$ point into the directions of the asymptotes. The tangent vector at point $\vec{p}(t)$ is:

$$\vec{p}'(t)=\vec{f_1}\sinh t+\vec{f_2}\cosh t.$$

Because at a vertex the tangent is perpendicular to the major axis of the hyperbola one gets the parameter t_0 of a vertex from the equation:

$$\vec{p}'(t)\cdot\left(\vec{p}(t)-\vec{f_0}\right)=\left(\vec{f_1}\sinh t+\vec{f_2}\cosh t\right)\cdot\left(\vec{f_1}\cosh t+\vec{f_2}\sinh t\right)=0$$

and hence from:

$$\coth(2t_0)=-\frac{\vec{f_1}^2+\vec{f_2}^2}{2\vec{f_1}\cdot\vec{f_2}},$$

which yields:

$$t_0=\tfrac{1}{4}\ln\frac{(\vec{f_1}-\vec{f_2})^2}{(\vec{f_1}+\vec{f_2})^2}.$$

(The formulae $\cosh^2 x+\sinh^2 x=\cosh 2x$, $2\sinh x\cosh x=\sinh 2x$, $\operatorname{arcoth} x=\tfrac{1}{2}\ln\frac{x+1}{x-1}$ were used.)

The two *vertices* of the hyperbola are $\vec{f}_0 \pm \left(\vec{f}_1 \cosh t_0 + \vec{f}_2 \sinh t_0 \right)$.

The advantage of this definition is that one gets a simple parametric representation of an arbitrary hyperbola, even in the space, if the vectors $\vec{f}_0, \vec{f}_1, \vec{f}_2$ are vectors of the Euclidean space.

Hyperbola as an Affine Image of the Hyperbola $y = 1/x$

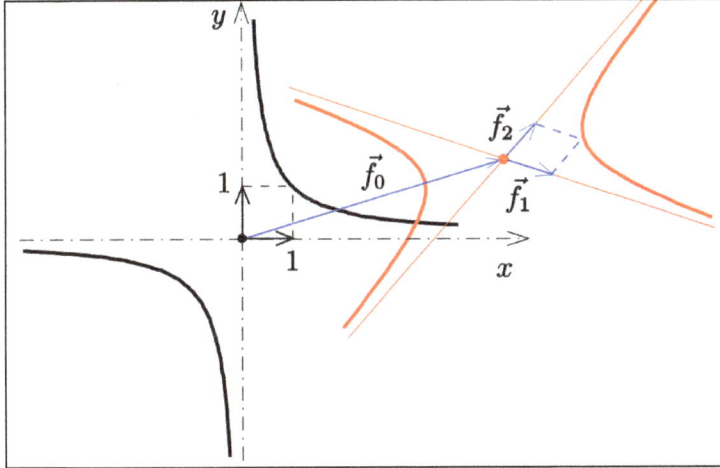

Hyperbola as affine image of $y = 1/x$.

Because the unit hyperbola $x^2 - y^2 = 1$ is affinely equivalent to the hyperbola $y = 1/x$, an arbitrary hyperbola can be considered as the affine image of the hyperbola $y = 1/x$:

$$\vec{x} = \vec{p}(t) = \vec{f}_0 + \vec{f}_1 t + \vec{f}_2 \tfrac{1}{t}, \quad t \neq 0.$$

$M : \vec{f}_0$ is the center of the hyperbola, the vectors $\vec{f}_1, \vec{f}_2$ have the directions of the asymptotes and $\vec{f}_1 + \vec{f}_2$ is a point of the hyperbola. The tangent vector is:

$$\vec{p}'(t) = \vec{f}_1 - \vec{f}_2 \tfrac{1}{t^2}.$$

At a vertex the tangent is perpendicular to the major axis. Hence:

$$\vec{p}'(t) \cdot \left(\vec{p}(t) - \vec{f}_0 \right) = \left(\vec{f}_1 - \vec{f}_2 \tfrac{1}{t^2} \right) \cdot \left(\vec{f}_1 t + \vec{f}_2 \tfrac{1}{t} \right) = \vec{f}_1^2 t - \vec{f}_2^2 \tfrac{1}{t^3} = 0$$

and the parameter of a vertex is

$$t_0 = \pm \sqrt[4]{\frac{\vec{f}_2^2}{\vec{f}_1^2}}.$$

$|\vec{f}_1| = |\vec{f}_2|$ is equivalent to $|\vec{f}_1| = |\vec{f}_2|$ and $\vec{f}_0 \pm (\vec{f}_1 + \vec{f}_2)$ are the vertices of the hyperbola.

The following properties of a hyperbola are easily proven using the representation of a hyperbola introduced in this part.

Tangent Construction

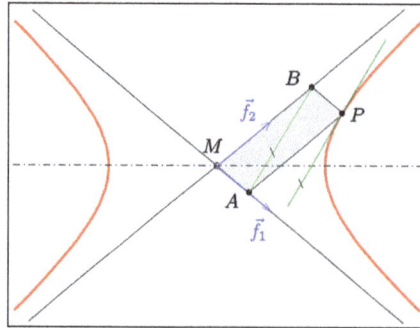

Tangent construction: asymptotes and P given → tangent.

The tangent vector can be rewritten by factorization:

$$\vec{p}'(t) = \tfrac{1}{t}\left(\vec{f_1} t - \vec{f_2}\tfrac{1}{t} \right).$$

This means that:

"The diagonal AB of the parallelogram $M: \vec{f_0}$, $A = \vec{f_0} + \vec{f_1}t$, $B: \vec{f_0} + \vec{f_2}\tfrac{1}{t}$, $P: \vec{f_0} + \vec{f_1}t + \vec{f_2}\tfrac{1}{t}$ is parallel to the tangent at the hyperbola point P".

- This property provides a way to construct the tangent at a point on the hyperbola.

- This property of a hyperbola is an affine version of the 3-point-degeneration of Pascal's theorem.

Area of the Grey Parallelogram

The area of the grey parallelogram $MAPB$ in the above diagram is:

$$\text{Area} = \left| \det\left(t\vec{f_1}, \tfrac{1}{t}\vec{f_2} \right) \right| = \left| \det\left(\vec{f_1}, \vec{f_2} \right) \right| = \cdots = \frac{a^2 + b^2}{4}$$

and hence independent of point P. The last equation follows from a calculation for the case, where P is a vertex and the hyperbola in its canonical form $\frac{x^2}{a^2} - \frac{y^2}{b^2} = 1$.

Point Construction

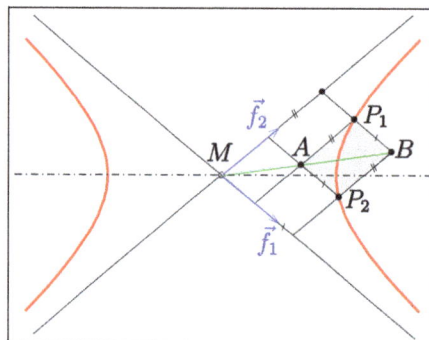

Point construction: asymptotes and P_1 are given → P_2.

For a hyperbola with parametric representation $\vec{x} = \vec{p}(t) = \vec{f_1}t + \vec{f_2}\frac{1}{t}$ (for simplicity the center is the origin) the following is true:

For any two points $P_1 : \vec{f_1}t_1 + \vec{f_2}\frac{1}{t_1}$, $P_2 : \vec{f_1}t_2 + \vec{f_2}\frac{1}{t_2}$ the points $A:\vec{a} = \vec{f_1}t_1 + \vec{f_2}\frac{1}{t_2}, B:\vec{b} = \vec{f_1}t_2 + \vec{f_2}\frac{1}{t_1}$ are collinear with the center of the hyperbola.

The simple proof is a consequence of the equation $\frac{1}{t_1}\vec{a} = \frac{1}{t_2}\vec{b}$.

This property provides a possibility to construct points of a hyperbola if the asymptotes and one point are given.

This property of a hyperbola is an affine version of the 4-point-degeneration of Pascal's theorem.

Tangent-asymptotes-triangle

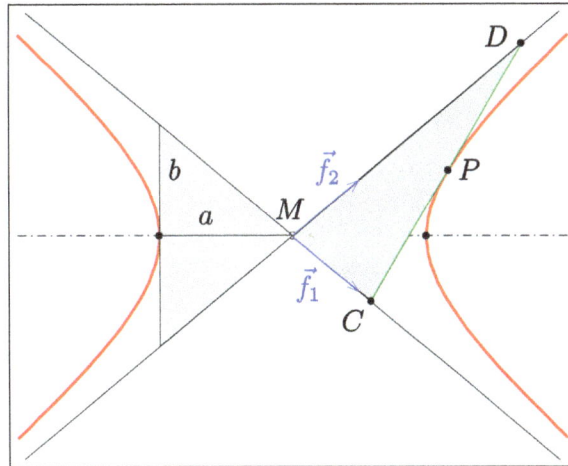

Hyperbola: tangent-asymptotes-triangle.

For simplicity the center of the hyperbola may be the origin and the vectors $\vec{f_1}, \vec{f_2}$ have equal length. If the last assumption is not fulfilled one can first apply a parameter transformation in order to make the assumption true. Hence $\pm(\vec{f_1} + \vec{f_2})$ are the vertices, $\pm(\vec{f_1} - \vec{f_2})$ span the minor axis and one gets $|\vec{f_1} + \vec{f_2}| = a$ and $|\vec{f_1} - \vec{f_2}| = b$.

For the intersection points of the tangent at point $\vec{p}(t_0) = \vec{f_1}t_0 + \vec{f_2}\frac{1}{t_0}$ with the asymptotes one gets the points:

$$C = 2t_0\vec{f_1}, D = \frac{2}{t_0}\vec{f_2}.$$

The *area* of the triangle M, C, D can be calculated by a 2×2-determinant:

$$A = \frac{1}{2}\left| \det\left(2t_0\vec{f_1}, \frac{2}{t_0}\vec{f_2}\right)\right| = 2\left|\det\left(\vec{f_1}, \vec{f_2}\right)\right|$$

$|\det(\vec{f_1}, \vec{f_2})|$ is the area of the rhombus generated by $\vec{f_1}, \vec{f_2}$. The area of a rhombus is equal to one half of the product of its diagonals. The diagonals are the semi-axes a, b of the hyperbola. Hence, The *area* of the triangle MCD is independent of the point of the hyperbola: $A = ab$.

Polar Coordinates

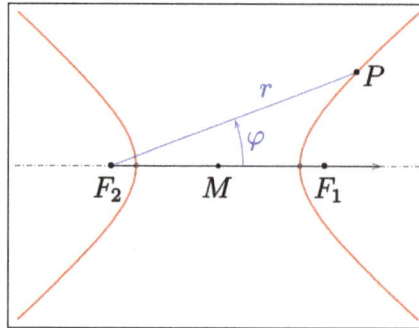

Hyperbola: Polar coordinates with pole = focus.

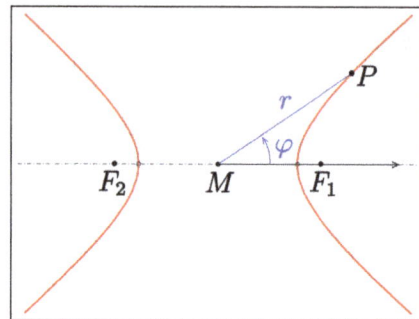

Hyperbola: Polar coordinates with pole = center.

For Pole = Focus

The polar coordinates used most commonly for the hyperbola are defined relative to the Cartesian coordinate system that has its *origin in a focus* and its x-axis pointing towards the origin of the "canonical coordinate system" as illustrated in the first diagram.

In this case the angle φ is called true anomaly.

Relative to this coordinate system one has that:

$$r = \frac{p}{1 \mp e \cos \varphi}, \quad p = \frac{b^2}{a}$$

and,

$$-\arccos\left(-\frac{1}{e}\right) < \varphi < \arccos\left(-\frac{1}{e}\right).$$

For Pole = Center

With polar coordinates relative to the "canonical coordinate system" one has that:

$$r = \frac{b}{\sqrt{e^2 \cos^2 \varphi - 1}}.$$

For the right branch of the hyperbola the range of φ is:

$$-\arccos\left(\frac{1}{e}\right) < \varphi < \arccos\left(\frac{1}{e}\right).$$

Parametric Equations

A hyperbola with equation $\frac{x^2}{a^2} - \frac{y^2}{b^2} = 1$ can be described by several parametric equations:

$$\begin{cases} x = \pm a\cosh t \\ y = b\sinh t \end{cases}, t \in \mathbb{R}.$$

$$\begin{cases} x = \pm a\frac{t^2+1}{2t} \\ y = b\frac{t^2-1}{2t} \end{cases}, t > 0. \ (rational \text{ representation})$$

$$\begin{cases} x = \dfrac{a}{\cos t} = a\sec t \\ y = \pm b\tan t \end{cases}, \quad 0 \le t < 2\pi; t \ne \frac{\pi}{2}; t \ne \frac{3}{2}\pi.$$

Tangent Slope as Parameter

A parametric representation, which uses the slope m of the tangent at a point of the hyperbola can be obtained analogously to the ellipse case: Replace in the ellipse case b^2 by $-b^2$ and use formulae for the hyperbolic functions. One gets:

$$\vec{c}_\pm(m) = \left(-\frac{ma^2}{\pm\sqrt{m^2a^2 - b^2}}, \frac{-b^2}{\pm\sqrt{m^2a^2 - b^2}}\right), |m| > b/a.$$

$\vec{c}_-$ is the upper and $\vec{c}_+$ the lower half of the hyperbola. The points with vertical tangents (vertices $(\pm a, 0)$) are not covered by the representation.

The equation of the tangent at point $\vec{c}_\pm(m)$ is:

$$y = mx \pm \sqrt{m^2a^2 - b^2}.$$

This description of the tangents of a hyperbola is an essential tool for the determination of the orthoptic of a hyperbola.

Reciprocation of a Circle

The reciprocation of a circle B in a circle C always yields a conic section such as a hyperbola. The process of "reciprocation in a circle C" consists of replacing every line and point in a geometrical figure with their corresponding pole and polar, respectively. The *pole* of a line is the inversion of its closest point to the circle C, whereas the polar of a point is the converse, namely, a line whose closest point to C is the inversion of the point.

The eccentricity of the conic section obtained by reciprocation is the ratio of the distances between the two circles' centers to the radius r of reciprocation circle C. If B and C represent the points at the centers of the corresponding circles, then:

$$e = \frac{\overline{BC}}{r}.$$

Since the eccentricity of a hyperbola is always greater than one, the center B must lie outside of the reciprocating circle C.

This definition implies that the hyperbola is both the locus of the poles of the tangent lines to the circle B, as well as the envelope of the polar lines of the points on B. Conversely, the circle B is the envelope of polars of points on the hyperbola, and the locus of poles of tangent lines to the hyperbola. Two tangent lines to B have no (finite) poles because they pass through the center C of the reciprocation circle C; the polars of the corresponding tangent points on B are the asymptotes of the hyperbola. The two branches of the hyperbola correspond to the two parts of the circle B that are separated by these tangent points.

Quadratic Equation

A hyperbola can also be defined as a second-degree equation in the Cartesian coordinates (x, y) in the plane,

$$A_{xx}x^2 + 2A_{xy}xy + A_{yy}y^2 + 2B_x x + 2B_y y + C = 0,$$

provided that the constants A_{xx}, A_{xy}, A_{yy}, B_x, B_y, and C satisfy the determinant condition:

$$D := \begin{vmatrix} A_{xx} & A_{xy} \\ A_{xy} & A_{yy} \end{vmatrix} < 0.$$

This determinant is conventionally called the discriminant of the conic section.

A special case of a hyperbola—the *degenerate hyperbola* consisting of two intersecting lines—occurs when another determinant is zero:

$$\Delta := \begin{vmatrix} A_{xx} & A_{xy} & B_x \\ A_{xy} & A_{yy} & B_y \\ B_x & B_y & C \end{vmatrix} = 0.$$

This determinant Δ is sometimes called the discriminant of the conic section.

Given the above general parametrization of the hyperbola in Cartesian coordinates, the eccentricity can be found using the formula in Conic section#Eccentricity in terms of parameters of the quadratic form.

The center (x_c, y_c) of the hyperbola may be determined from the formulae:

$$y_c = -\frac{1}{D}\begin{vmatrix} A_{xx} & B_x \\ A_{xy} & B_y \end{vmatrix}.$$

In terms of new coordinates, $\xi = x - x_c$ and $\eta = y - y_c$, the defining equation of the hyperbola can be written:

$$A_{xx}\xi^2 + 2A_{xy}\xi\eta + A_{yy}\eta^2 + \frac{\Delta}{D} = 0.$$

The principal axes of the hyperbola make an angle φ with the positive x-axis that is given by:

$$\tan 2\varphi = \frac{2A_{xy}}{A_{xx} - A_{yy}}.$$

Rotating the coordinate axes so that the x-axis is aligned with the transverse axis brings the equation into its canonical form:

$$\frac{x^2}{a^2} - \frac{y^2}{b^2} = 1.$$

The major and minor semiaxes a and b are defined by the equations:

$$a^2 = -\frac{\Delta}{\lambda_1 D} = -\frac{\Delta}{\lambda_1^2 \lambda_2},$$

$$b^2 = -\frac{\Delta}{\lambda_2 D} = -\frac{\Delta}{\lambda_1 \lambda_2^2},$$

where λ_1 and λ_2 are the roots of the quadratic equation:

$$\lambda^2 - \left(A_{xx} + A_{yy}\right)\lambda + D = 0.$$

For comparison, the corresponding equation for a degenerate hyperbola (consisting of two intersecting lines) is:

$$\frac{x^2}{a^2} - \frac{y^2}{b^2} = 0.$$

The tangent line to a given point (x_0, y_0) on the hyperbola is defined by the equation:

$$Ex + Fy + G = 0$$

where E, F and G are defined by:

$$E = A_{xx}x_0 + A_{xy}y_0 + B_x,$$

$$F = A_{xy}x_0 + A_{yy}y_0 + B_y,$$

$$G = B_xx_0 + B_yy_0 + C.$$

The normal line to the hyperbola at the same point is given by the equation:

$$F(x-x_0) - E(y-y_0) = 0.$$

The normal line is perpendicular to the tangent line, and both pass through the same point (x_0, y_0).

From the equation:

$$\frac{x^2}{a^2} - \frac{y^2}{b^2} = 1, \qquad 0 < b \le a,$$

the left focus is $(-ae, 0)$ and the right focus is $(ae, 0)$, where e is the eccentricity. Denote the distances from a point (x, y) to the left and right foci as r_1 and r_2. For a point on the right branch,

$$r_1 - r_2 = 2a,$$

and for a point on the left branch,

$$r_2 - r_1 = 2a.$$

This can be proved as follows:

1. If (x,y) is a point on the hyperbola the distance to the left focal point is:

$$r_1^2 = (x+ae)^2 + y^2 = x^2 + 2xae + a^2e^2 + (x^2 - a^2)(e^2 - 1) = (ex+a)^2.$$

To the right focal point the distance is:

$$r_2^2 = (x-ae)^2 + y^2 = x^2 - 2xae + a^2e^2 + (x^2 - a^2)(e^2 - 1) = (ex-a)^2.$$

2. If (x,y) is a point on the right branch of the hyperbola then $ex > a$ and:

$$r_1 = ex + a,$$

$$r_2 = ex - a.$$

Subtracting these equations one gets:

$$r_1 - r_2 = 2a.$$

3. If (x,y) is a point on the left branch of the hyperbola then $ex < -a$ and:

$$r_1 = -ex - a,$$

$$r_2 = -ex + a.$$

Subtracting these equations one gets:

$$r_2 - r_1 = 2a.$$

Conic Section Analysis of the Hyperbolic Appearance of Circles

Besides providing a uniform description of circles, ellipses, parabolas, and hyperbolas, conic sections can also be understood as a natural model of the geometry of perspective in the case where the scene being viewed consists of circles, or more generally an ellipse. The viewer is typically a camera or the human eye and the image of the scene a central projection onto an image plane, that is, all projection rays pass a fixed point O, the center. The lens plane is a plane parallel to the image plane at the lens O.

The image of a circle c is:

- A circle, if circle c is in a special position, for example parallel to the image plane and others.

- An ellipse, if c has *no* point with the lens plane in common.

- A parabola, if c has *one* point with the lens plane in common.

- A hyperbola, if c has *two* points with the lens plane in common.

(Special positions where the circle plane contains point O are omitted.)

These results can be understood if one recognizes that the projection process can be seen in two steps: 1) circle c and point O generate a cone which is 2) cut by the image plane, in order to generate the image.

One sees a hyperbola whenever catching sight of a portion of a circle cut by one's lens plane. The inability to see very much of the arms of the visible branch, combined with the complete absence of the second branch, makes it virtually impossible for the human visual system to recognize the connection with hyperbolas.

Arc Length

The arc length of a hyperbola does not have a closed-form expression. The upper half of a hyperbola can be parameterized as:

$$y = b\sqrt{\frac{x^2}{a^2} - 1}.$$

Then the integral giving the arc length s from x_1 to x_2 can be computed numerically,

$$s = b \int_{\text{arcosh}\frac{x_1}{a}}^{\text{arcosh}\frac{x_2}{a}} \sqrt{1 + \left(1 + \frac{a^2}{b^2}\right)\sinh^2 v} \, dv$$

After using the substitution $z = iv$, this can also be represented using the elliptic integral of the second kind with parameter $m = k^2$:

$$s = -ib\left[E\left(iz \middle| 1 + \frac{a^2}{b^2} \right) \right]_{\text{arcosh}\frac{x_1}{a}}^{\text{arcosh}\frac{x_2}{a}}$$

Derived Curves

Several other curves can be derived from the hyperbola by inversion, the so-called inverse curves of the hyperbola. If the center of inversion is chosen as the hyperbola's own center, the inverse curve is the lemniscate of Bernoulli; the lemniscate is also the envelope of circles centered on a rectangular hyperbola and passing through the origin. If the center of inversion is chosen at a focus or a vertex of the hyperbola, the resulting inverse curves are a limaçon or a strophoid, respectively.

Elliptic Coordinates

A family of confocal hyperbolas is the basis of the system of elliptic coordinates in two dimensions. These hyperbolas are described by the equation

$$\left(\frac{x}{c\cos\theta}\right)^2 - \left(\frac{y}{c\sin\theta}\right)^2 = 1$$

where the foci are located at a distance c from the origin on the x-axis, and where θ is the angle of the asymptotes with the x-axis. Every hyperbola in this family is orthogonal to every ellipse that shares the same foci. This orthogonality may be shown by a conformal map of the Cartesian coordinate system $w = z + 1/z$, where $z = x + iy$ are the original Cartesian coordinates, and $w = u + iv$ are those after the transformation.

Other orthogonal two-dimensional coordinate systems involving hyperbolas may be obtained by other conformal mappings. For example, the mapping $w = z^2$ transforms the Cartesian coordinate system into two families of orthogonal hyperbolas.

Other Properties of Hyperbolas

- The following are concurrent: (1) a circle passing through the hyperbola's foci and centered at the hyperbola's center; (2) either of the lines that are tangent to the hyperbola at the vertices; and (3) either of the asymptotes of the hyperbola.

- The following are also concurrent: (1) the circle that is centered at the hyperbola's center and that passes through the hyperbola's vertices; (2) either directrix; and (3) either of the asymptotes.

Applications

Hyperbolas as declination lines on a sundial.

Sundials

Hyperbolas may be seen in many sundials. On any given day, the sun revolves in a circle on the celestial sphere, and its rays striking the point on a sundial traces out a cone of light. The intersection of this cone with the horizontal plane of the ground forms a conic section. At most populated latitudes and at most times of the year, this conic section is a hyperbola. In practical terms, the shadow of the tip of a pole traces out a hyperbola on the ground over the course of a day (this path is called the declination line). The shape of this hyperbola varies with the geographical latitude and with the time of the year, since those factors affect the cone of the sun's rays relative to the horizon. The collection of such hyperbolas for a whole year at a given location was called a *pelekinon* by the Greeks, since it resembles a double-bladed axe.

Multilateration

A hyperbola is the basis for solving multilateration problems, the task of locating a point from the differences in its distances to given points — or, equivalently, the difference in arrival times of synchronized signals between the point and the given points. Such problems are important in navigation, particularly on water; a ship can locate its position from the difference in arrival times of signals from a LORAN or GPS transmitters. Conversely, a homing beacon or any transmitter can be located by comparing the arrival times of its signals at two separate receiving stations; such techniques may be used to track objects and people. In particular, the set of possible positions of a point that has a distance difference of $2a$ from two given points is a hyperbola of vertex separation $2a$ whose foci are the two given points.

Path followed by a Particle

The path followed by any particle in the classical Kepler problem is a conic section. In particular, if the total energy E of the particle is greater than zero (that is, if the particle is unbound), the path of such a particle is a hyperbola. This property is useful in studying atomic and sub-atomic forces by scattering high-energy particles; for example, the Rutherford experiment demonstrated the existence of an atomic nucleus by examining the scattering of alpha particles from gold atoms. If the short-range nuclear interactions are ignored, the atomic nucleus and the alpha particle interact only by a repulsive Coulomb force, which satisfies the inverse square law requirement for a Kepler problem.

Korteweg–de Vries Equation

The hyperbolic trig function sech x appears as one solution to the Korteweg–de Vries equation which describes the motion of a soliton wave in a canal.

Angle Trisection

As shown first by Apollonius of Perga, a hyperbola can be used to trisect any angle, a well studied problem of geometry. Given an angle, first draw a circle centered at its vertex O, which intersects the sides of the angle at points A and B. Next draw the line segment with endpoints A and B and its perpendicular bisector ℓ. Construct a hyperbola of eccentricity $e=2$ with ℓ as directrix and B as a focus. Let P be the intersection (upper) of the hyperbola with the circle. Angle POB trisects angle AOB.

To prove this, reflect the line segment OP about the line ℓ obtaining the point P' as the image of P. Segment AP' has the same length as segment BP due to the reflection, while segment PP' has the same length as segment BP due to the eccentricity of the hyperbola. As OA, OP', OP and OB are all radii of the same circle (and so, have the same length), the triangles OAP', OPP' and OPB are all congruent. Therefore, the angle has been trisected, since 3×POB = AOB.

Efficient Portfolio Frontier

In portfolio theory, the locus of mean-variance efficient portfolios (called the efficient frontier) is the upper half of the east-opening branch of a hyperbola drawn with the portfolio return's standard deviation plotted horizontally and its expected value plotted vertically; according to this theory, all rational investors would choose a portfolio characterized by some point on this locus.

Biochemistry

In biochemistry and pharmacology, the Hill equation and Hill-Langmuir equation respectively describe biological responses and the formation of protein–ligand complexes as functions of ligand concentration. They are both rectangular hyperbolae.

PARABOLA

One description of a parabola involves a point (the focus) and a line (the directrix). The focus does not lie on the directrix. The parabola is the locus of points in that plane that are equidistant from both the directrix and the focus. Another description of a parabola is as a conic section, created from the intersection of a right circular conical surface and a plane which is parallel to another plane that is tangential to the conical surface.

The line perpendicular to the directrix and passing through the focus (that is, the line that splits the parabola through the middle) is called the "axis of symmetry". The point on the parabola that intersects the axis of symmetry is called the "vertex", and is the point where the parabola is most sharply curved. The distance between the vertex and the focus, measured along the axis of symmetry, is the "focal length". The "latus rectum" is the chord of the parabola which is parallel to the directrix and passes through the focus. Parabolas can open up, down, left, right, or in some other arbitrary direction. Any parabola can be repositioned and rescaled to fit exactly on any other parabola—that is, all parabolas are geometrically similar.

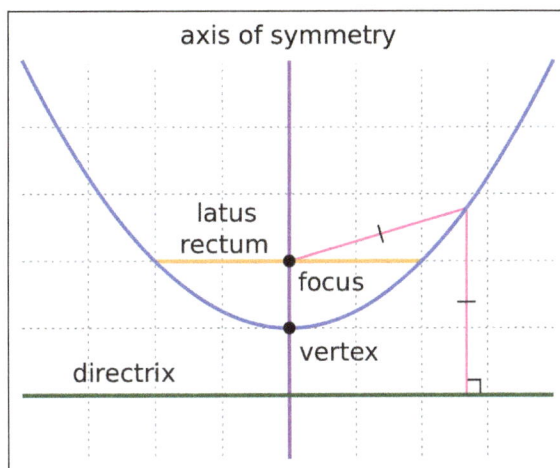

Part of a parabola (blue), with various features (other colours). The complete parabola has no endpoints. In this orientation, it extends infinitely to the left, right, and upward.

Parabolas have the property that, if they are made of material that reflects light, then light which travels parallel to the axis of symmetry of a parabola and strikes its concave side is reflected to its focus, regardless of where on the parabola the reflection occurs. Conversely, light that originates from a point source at the focus is reflected into a parallel ("collimated") beam, leaving the parabola parallel to the axis of symmetry. The same effects occur with sound and other forms of energy. This reflective property is the basis of many practical uses of parabolas.

The parabola has many important applications, from a parabolic antenna or parabolic microphone to automobile headlight reflectors to the design of ballistic missiles. They are frequently used in physics, engineering, and many other areas.

Locus of Points

A parabola can be defined geometrically as a set of points (locus of points) in the Euclidean plane:

- A parabola is a set of points, such that for any point P of the set the distance $|PF|$ to a fixed point F, the *focus*, is equal to the distance $|Pl|$ to a fixed line l, the *directrix*:

$$\{P : |PF| = |Pl|\}$$

The midpoint V of the perpendicular from the focus F onto the directrix is called *vertex* and the line FV the *axis of symmetry* of the parabola.

In a Cartesian Coordinate System

Axis of Symmetry Parallel to the y-axis

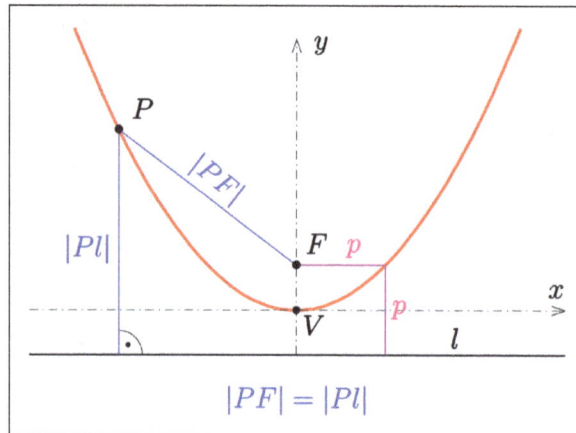

Parabola: Definition, p: *semi-latus rectum*.

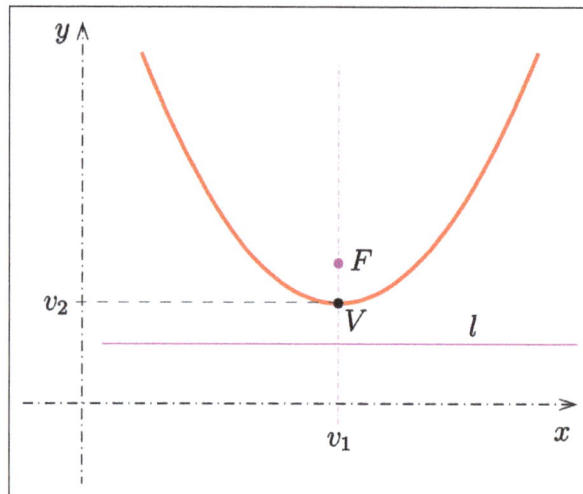

Parabola: axis parallel to y-axis.

If one introduces cartesian coordinates, such that $F = (0, f), f > 0$, and the directrix has the equation $y = -f$ one obtains for a point $P = (x, y)$ from $|PF|^2 = |Pl|^2$ the equation $x^2 + (y - f)^2 = (y + f)^2$. Solving for y yields:

$$y = \frac{1}{4f} x^2$$

The horizontal chord through the focus is called the *latus rectum*; one half of it is the *semi-latus rectum*. The latus rectum is parallel to the directrix. The semi-latus rectum is designated by the letter p. From the picture one obtains:

$$p = 2f.$$

The latus rectum is defined similarly for the other two conics, namely the ellipse and the hyperbola, respectively. The latus rectum is the line drawn through a focus of a conic section parallel to the directrix and terminated both ways by the curve. For any case, p is the radius of the osculating circle at the vertex. For a parabola, the semi-latus rectum, p, is the distance of the focus from the directrix. Using the parameter p, the equation of the parabola can be rewritten as:

$$x^2 = 2py$$

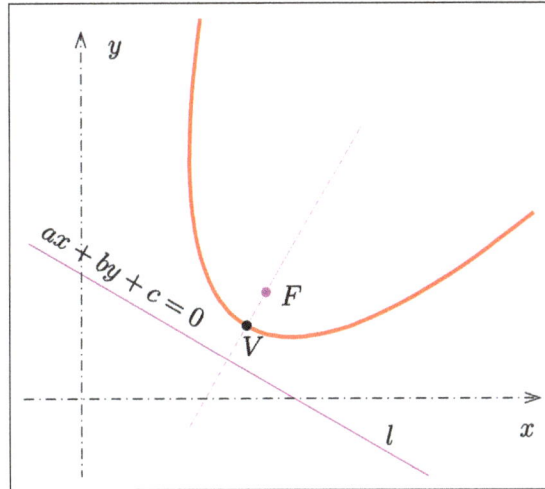

Parabola: general case.

More generally, if the vertex is $V = (v_1, v_2)$, the focus $F = (v_1, v_2 + f)$ and the directrix $y = v_2 - f$, one obtains the equation:

$$y = \frac{1}{4f}(x - v_1)^2 + v_2 = \frac{1}{4f}x^2 - \frac{v_1}{2f}x + \frac{v_1^2}{4f} + v_2 .$$

1. In the case of $f < 0$ the parabola has a downwards opening.

2. The presumption that the *axis is parallel to the y-axis* allows one to consider a parabola as the graph of a polynomial of degree 2, and vice versa: the graph of an arbitrary polynomial of degree 2 is a parabola.

3. If one exchanges x and y, one obtains equations of the form $y^2 = 2px$. These parabolas open to the left (if $p < 0$) or to the right (if $p > 0$).

If the focus is $F = (f_1, f_2)$ and the directrix $ax + by + c = 0$ one obtains the equation:

$$\frac{(ax + by + c)^2}{a^2 + b^2} = (x - f_1)^2 + (y - f_2)^2$$

(The left side of the equation uses the Hesse normal form of a line to calculate the distance $|Pl|$.)

For a parametric equation of a parabola in general position see § As the affine image of the unit parabola.

The implicit equation of a parabola is defined by an irreducible polynomial of degree two:

$$ax^2 + bxy + cy^2 + dx + ey + f = 0,$$

such that $b^2 - 4ac = 0$, or, equivalently, such that $ax^2 + bxy + cy^2$ is the square of a linear polynomial.

As a Graph of a Function

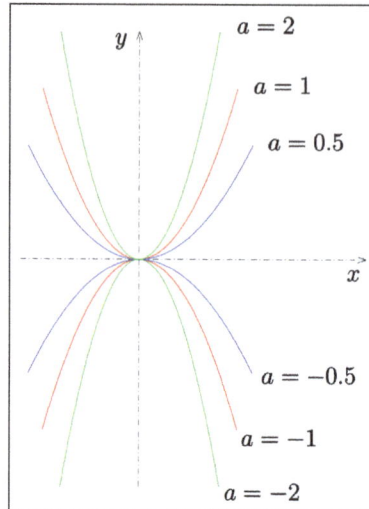

Parabolas $y = ax^2$.

The previous section shows: any parabola with the origin as vertex and the y-axis as axis of symmetry can be considered as the graph of a function:

$$f(x) = ax^2 \text{ with } a \neq 0.$$

For $a > 0$ the parabolas are opening to the top and for $a < 0$ opening to the bottom. From the section above one obtains:

- The *focus* is $\left(0, \dfrac{1}{4a}\right)$,
- The *focal length* $\dfrac{1}{4a}$, the *semi-latus rectum* is $p = \dfrac{1}{2a}$,
- The *vertex* is $(0,0)$,
- The *directrix* has the equation $y = -\dfrac{1}{4a}$,
- The *tangent* at point $\left(x_0, ax_0^2\right)$ has the equation $y = 2ax_0 x - ax_0^2$.

For $a = 1$ the parabola is the unit parabola with equation $y = x^2$. Its focus is $\left(0, \frac{1}{4}\right)$, the semi-latus rectum $p = \frac{1}{2}$ and the directrix has the equation $y = -\frac{1}{4}$.

The general function of degree 2 is:

$$f(x) = ax^2 + bx + c \text{ with } a, b, c \in \mathbb{R}, a \neq 0.$$

Completing the square yields:

$$f(x) = a\left(x + \frac{b}{2a}\right)^2 + \frac{4ac - b^2}{4a},$$

which is the equation of a parabola with:

- The axis $x = -\dfrac{b}{2a}$ (parallel to the y-axis),

- The *focal length* $\dfrac{1}{4a}$, the *semi-latus rectum* $p = \dfrac{1}{2a}$,

- The *vertex* $V = \left(-\dfrac{b}{2a}, \dfrac{4ac - b^2}{4a}\right),$

- The *focus* $F = \left(-\dfrac{b}{2a}, \dfrac{4ac - b^2 + 1}{4a}\right),$

- The *directrix* $y = \dfrac{4ac - b^2 - 1}{4a},$

- The point of the parabola intersecting the y-axis has coordinates $(0, c)$,

- The *tangent* at a point on the y-axis has the equation $y = bx + c$.

Similarity to the Unit Parabola

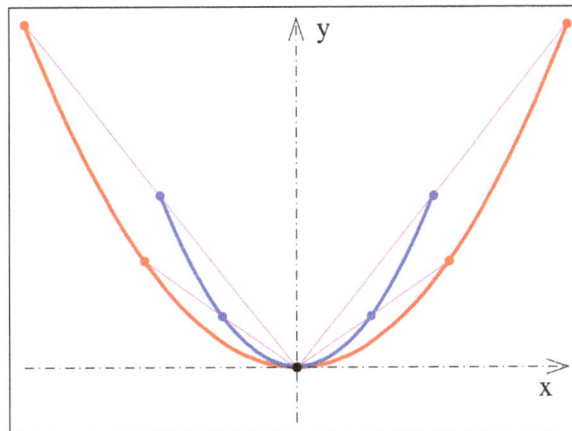

When the parabola $y = 2x^2$ is uniformly scaled by factor 2, the result is the parabola $y = x^2$.

Two objects in the Euclidean plane are *similar* if one can be transformed to the other by a *similarity*, that is, an arbitrary composition of rigid motions (translations and rotations) and uniform scalings.

A parabola $\mathcal{P}$ with vertex $V = (v_1, v_2)$ can be transformed by the translation $(x, y) \to (x - v_1, y - v_2)$ to one with the origin as vertex. A suitable rotation around the origin can then transform the parabola to one that has the y-axis as axis of symmetry. Hence the parabola $\mathcal{P}$ can be transformed by a rigid motion to a parabola with an equation $y = ax^2, a \neq 0$. Such a parabola can then be transformed by the uniform scaling $(x,y) \to (ax, ay)$ into the unit parabola with equation $y = x^2$. Thus, any parabola can be mapped to the unit parabola by a similarity.

A synthetic approach, using similar triangles, can also be used to establish this result.

The general result is that two conic sections (necessarily of the same type) are similar if and only if they have the same eccentricity. Thus, only circles (all having eccentricity 0) share this property with parabolas (all having eccentricity 1), while general ellipses and hyperbolas do not.

There are other simple affine transformations that map the parabola $y = ax^2$ onto the unit parabola, such as $(x,y) \to \left(x, \frac{y}{a}\right)$. But this mapping is not a similarity, and only shows that all parabolas are affinely equivalent.

As a Special Conic Section

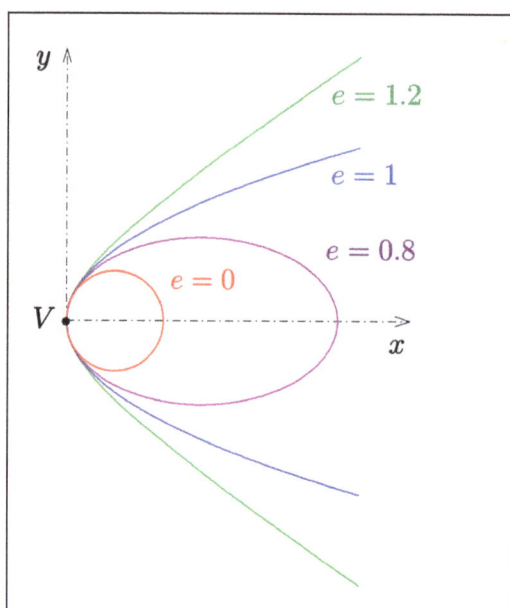

Pencil of conics with a common vertex

The pencil of conic sections with the x-axis as axis of symmetry, one vertex at the origin (0,0) and the same semi-latus rectum p can be represented by the equation:

$$y^2 = 2px + (e^2 - 1)x^2 \quad , e \geq 0,$$

with e the eccentricity.

- For $e = 0$ the conic is a *circle* (osculating circle of the pencil),

- for $0 < e < 1$ an *ellipse,*

- for $e = 1$ the parabola with equation $y^2 = 2px$ and

- for $e > 1$ a hyperbola.

In Polar Coordinates

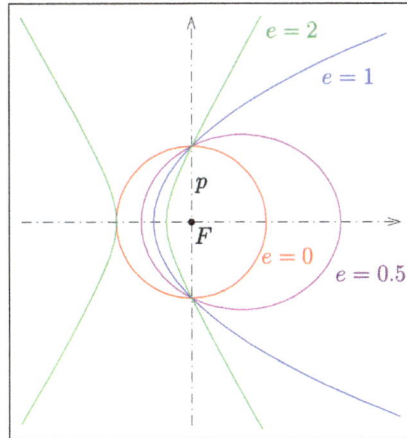

Pencil of conics with a common focus.

If $p > 0$, the parabola with equation $y^2 = 2px$ (opening to the right) has the polar coordinate representation:

$$r = 2p\,\frac{\cos\varphi}{\sin^2\varphi} \text{ with } \varphi \in \left[-\frac{\pi}{2},\frac{\pi}{2}\right] \setminus \{0\}.$$

$$r^2 = x^2 + y^2, \, x = r\cos\varphi$$

Its vertex is $V = (0,0)$ and its focus is $F = \left(\frac{p}{2},0\right)$.

If one shifts the origin into the focus, i.e., $F = (0,0)$, one obtains the equation:

$$r = \frac{p}{1-\cos\varphi} \text{ with } \varphi \neq 2\pi k.$$

Inverting this polar form shows: a parabola is the inverse of a cardioid.

The second polar form is a special case of a pencil of conics with focus $F = (0,0)$:

$$r = \frac{p}{1-e\cos\varphi}, \, (e\text{ : eccentricity}).$$

Conic Section and Quadratic Form

The diagram represents a cone with its axis vertical. The point A is its apex. An inclined cross-section of the cone, shown in pink, is inclined from the vertical by the same angle, θ, as the side of the cone. According to the definition of a parabola as a conic section, the boundary of this pink cross-section, EPD, is a parabola.

A horizontal cross-section of the cone passes through the vertex, P, of the parabola. This cross-section is circular, but appears elliptical when viewed obliquely, as is shown in the diagram. Its centre is V, and PK is a diameter. We will call its radius r.

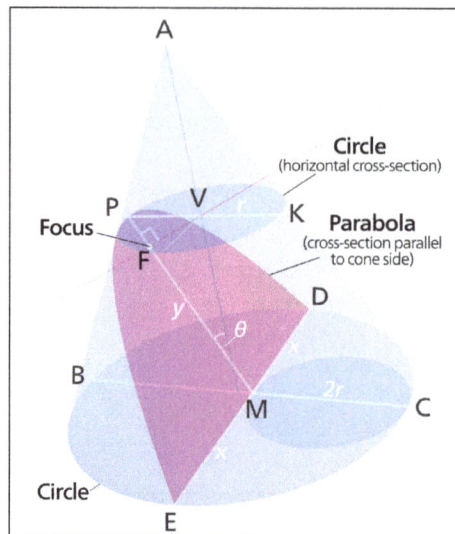

Cone with cross-sections.

Another horizontal, circular cross-section of the cone is farther from the apex, A, than the one just described. It has a chord DE, which joins the points where the parabola intersects the circle. Another chord, BC, is the perpendicular bisector of DE, and is consequently a diameter of the circle. These two chords and the parabola's axis of symmetry, PM, all intersect at the point M.

All the labelled points, except D and E, are coplanar. They are in the plane of symmetry of the whole figure. This includes the point F, which is not mentioned above. It is defined and discussed below, in the paragraph "Position of the focus".

Let us call the length of DM and of EM x, and the length of PM y.

Derivation of Quadratic Equation

The lengths of BM and CM are:

$$\overline{BM} = 2y\sin\theta \text{ (triangle BPM is isosceles.)}$$

$$\overline{CM} = 2r \text{ (PMCK is a parallelogram.)}$$

Using the intersecting chords theorem on the chords BC and DE, we get:

$$\overline{BM}\cdot\overline{CM} = \overline{DM}\cdot\overline{EM}$$

Substituting:

$$4ry\sin\theta = x^2$$

Rearranging:

$$y = \frac{x^2}{4r\sin\theta}$$

For any given cone and parabola, r and θ are constants, but x and y are variables which depend on the arbitrary height at which the horizontal cross-section BECD is made. This last equation shows the relationship between these variables. They can be interpreted as Cartesian coordinates of the points D and E, in a system in the pink plane with P as its origin. Since x is squared in the equation, the fact that D and E are on opposite sides of the y-axis is unimportant. If the horizontal cross-section moves up or down, toward or away from the apex of the cone, D and E move along the parabola, always maintaining the relationship between x and y shown in the equation. The parabolic curve is therefore the locus of points where the equation is satisfied, which makes it a Cartesian graph of the quadratic function in the equation.

This discussion started from the definition of a parabola as a conic section, but it has now led to a description as a graph of a quadratic function. This shows that these two descriptions are equivalent. They both define curves of exactly the same shape.

Focal Length

It is proved in a preceding section that if a parabola has its vertex at the origin, and if it opens in the positive y direction, then its equation is $y = \dfrac{x^2}{4f}$, where f is its focal length.[c] Comparing this with the last equation above shows that the focal length of the parabola in the cone is $r \sin \theta$.

Position of the Focus

In the diagram above, the point V is the foot of the perpendicular from the vertex of the parabola to the axis of the cone. The point F is the foot of the perpendicular from the point V to the plane of the parabola. By symmetry, F is on the axis of symmetry of the parabola. Angle VPF is complementary to θ, and angle PVF is complementary to angle VPF, therefore angle PVF is θ. Since the length of $\overline{PV}$ is r, the distance of F from the vertex of the parabola is $r \sin \theta$. It is shown above that this distance equals the focal length of the parabola, which is the distance from the vertex to the focus. The focus and the point F are therefore equally distant from the vertex, along the same line, which implies that they are the same point. Therefore, the point F, defined above, is the focus of the parabola.

Alternative Proof with Dandelin Spheres

An alternative proof can be done using Dandelin spheres. It works without calculation and uses elementary geometric considerations, only.

Proof of the Reflective Property

The reflective property states that, if a parabola can reflect light, then light which enters it travelling parallel to the axis of symmetry is reflected toward the focus. In the following proof, the fact that every point on the parabola is equidistant from the focus and from the directrix is taken as axiomatic.

Consider the parabola $y = x^2$. Since all parabolas are similar, this simple case represents all others. The right-hand side of the diagram shows part of this parabola.

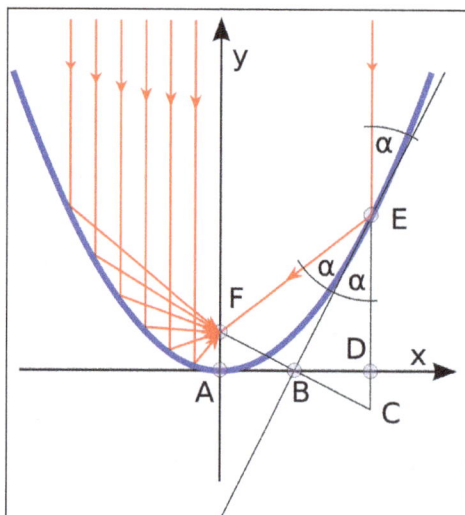

Reflective property of a parabola.

Construction and Definitions

The point E is an arbitrary point on the parabola, with coordinates (x, x^2). The focus is F, the vertex is A (the origin), and the line $\overline{FA}$ (the y-axis) is the axis of symmetry. The line $\overline{EC}$ is parallel to the axis of symmetry, and intersects the x-axis at D. The point C is located on the directrix (which is not shown, to minimize clutter). The point B is the midpoint of the line segment $\overline{FC}$.

Deductions

Measured along the axis of symmetry, the vertex, A, is equidistant from the focus, F, and from the directrix. According to the Intercept theorem, since C is on the directrix, the y-coordinates of F and C are equal in absolute value and opposite in sign. B is the midpoint of FC, so its y-coordinate is zero, so it lies on the x-axis. Its x-coordinate is half that of E, D, and C, i.e., $\frac{x}{2}$. The slope of the line $\overline{BE}$ is the quotient of the lengths of $\overline{ED}$ and $\overline{BD}$, which is $\frac{x^2}{\frac{x}{2}}$, which comes to 2x. But 2x is also the slope (first derivative) of the parabola at E.

Therefore, the line $\overline{BE}$ is the tangent to the parabola at E.

The distances $\overline{EF}$ and $\overline{EC}$ are equal because E is on the parabola, F is the focus and C is on the directrix. Therefore, since B is the midpoint of $\overline{FC}$, triangles △FEB and △CEB are congruent (three sides), which implies that the angles marked α are congruent. (The angle above E is vertically opposite angle ∠BEC.) This means that a ray of light which enters the parabola and arrives at E travelling parallel to the axis of symmetry will be reflected by the line $\overline{BE}$ so it travels along the line $\overline{EF}$, as shown in red in the diagram (assuming that the lines can somehow reflect light). Since $\overline{BE}$ is the tangent to the parabola at E, the same reflection will be done by an infinitesimal arc of the parabola at E. Therefore, light that enters the parabola and arrives at E travelling parallel to the axis of symmetry of the parabola is reflected by the parabola toward its focus.

The point E has no special characteristics. This conclusion about reflected light applies to all points on the parabola, as is shown on the left side of the diagram. This is the reflective property.

Tangent Bisection Property

The above proof and the accompanying diagram show that the tangent BE bisects the angle ∠FEC. In other words, the tangent to the parabola at any point bisects the angle between the lines joining the point to the focus, and perpendicularly to the directrix.

Intersection of a Tangent and Perpendicular from Focus

Since triangles △FBE and △CBE are congruent, FB is perpendicular to the tangent BE. Since B is on the x-axis, which is the tangent to the parabola at its vertex, it follows that the point of intersection between any tangent to a parabola and the perpendicular from the focus to that tangent lies on the line that is tangential to the parabola at its vertex.

Reflection of Light Striking the Convex Side

If light travels along the line $\overline{CE}$, it moves parallel to the axis of symmetry and strikes the convex side of the parabola at E. It is clear from the above diagram that this light will be reflected directly away from the focus, along an extension of the segment $\overline{FE}$.

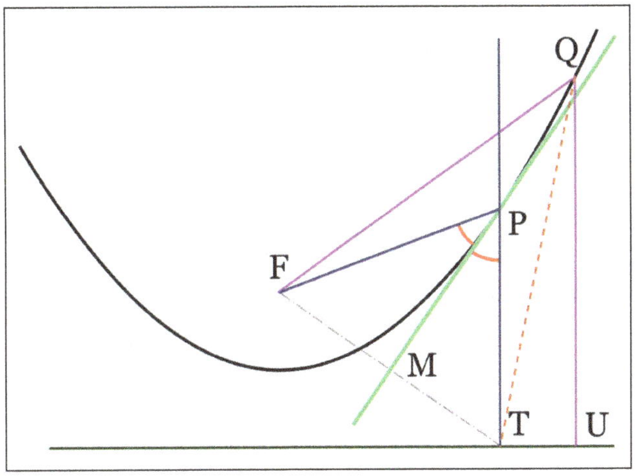

Parabola and tangent.

The above proofs of the reflective and tangent bisection properties use a line of calculus. For readers who are not comfortable with calculus, the following alternative is presented.

In this diagram, F is the focus of the parabola, and T and U lie on its directrix. P is an arbitrary point on the parabola. $\overline{PT}$ is perpendicular to the directrix, and the line $\overline{MP}$ bisects angle ∠FPT. Q is another point on the parabola, with $\overline{QU}$ perpendicular to the directrix. We know that $\overline{FP} = \overline{PT}$ and $\overline{FQ} = \overline{QU}$. Clearly, $\overline{QT} > \overline{QU}$, so $\overline{QT} > \overline{FQ}$. All points on the bisector MP are equidistant from F and T, but Q is closer to F than to T. This means that Q is to the left of MP, i.e., on the same side of it as the focus. The same would be true if Q were located anywhere else on the parabola (except at the point P), so the entire parabola, except the point P, is on the focus side of $\overline{MP}$.

Therefore, $\overline{MP}$ is the tangent to the parabola at P. Since it bisects the angle ∠FPT, this proves the tangent bisection property.

The logic of the last paragraph can be applied to modify the above proof of the reflective property. It effectively proves the line BE to be the tangent to the parabola at E if the angles α are equal. The reflective property follows as shown.

Pin and String Construction

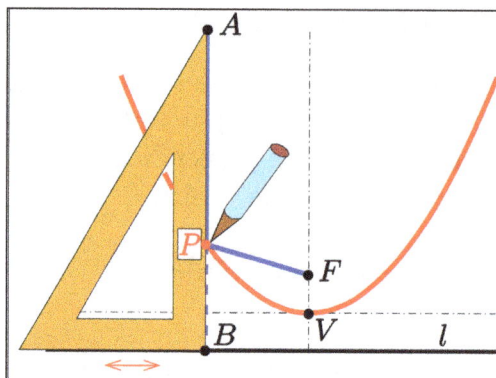

Parabola: pin string construction.

The definition of a parabola by its focus and directrix can be used for drawing it with help of pins and strings:

- Choose the *focus F* and the *directrix l* of the parabola.

- Take a triangle of a *set square* and prepare a *string* with length $|AB|$.

- Pin one end of the string at point A of the triangle and the other one to the focus F.

- Position the triangle such that the second edge of the right angle is free to *slide* along the directrix.

- Take a *pen* and hold the string tight to the triangle.

- While moving the triangle along the directrix the pen *draws* an arc of a parabola, because of $|PF|=|PB|$.

Properties Related to Pascal's Theorem

A parabola can be considered as the affine part of a non degenerated projective conic with a point Y_∞ on the line of infinity g_∞, which is the tangent at Y_∞. The 5-,4- and 3- point degenerations of Pascal's theorem are properties of a conic dealing with at least one tangent. If one considers this tangent as the line at infinity and its point of contact as the point at infinity of the y-axis, one obtains three statements for a parabola.

The following properties of a parabola deal only with terms *connect, intersect, parallel*, which are invariants of similarities. So, it is sufficient to prove any property for the *unit parabola* with equation $y = x^2$.

4-points-property

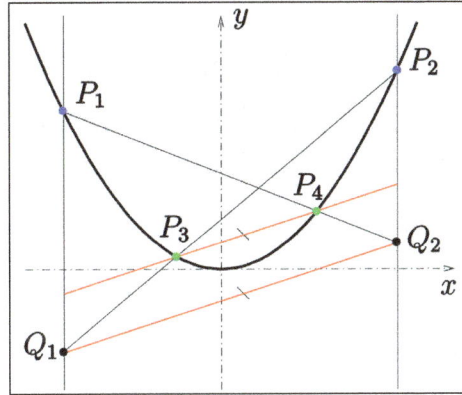

4-points-property of a parabola.

Any parabola can be described in a suitable coordinate system by an equation $y = ax^2$.

- Let $P_1 = (x_1, y_1), P_2 = (x_2, y_2), P_3 = (x_3, y_3), P_4 = (x_4, y_4)$ be four points of the parabola $y = ax^2$ and Q_2 the intersection of the secant line $P_1 P_4$ with the line $x = x_2$, and let Q_1 be the intersection of the secant line $P_2 P_3$ with the line $x = x_1$

- Then the secant line $P_3 P_4$ is parallel to line $Q_1 Q_2$.

  (The lines $x = x_1$ and $x = x_2$ are parallel to the axis of the parabola.)

Proof: straightforward calculation for the unit parabola $y = x^2$.

Application: The 4-points-property of a parabola can be used for the construction of point P_4, while P_1, P_2, P_3 and Q_2 are given.

The 4-points-property of a parabola is an affine version of the 5-point-degeneration of Pascal's theorem.

3-points-1-tangent-property

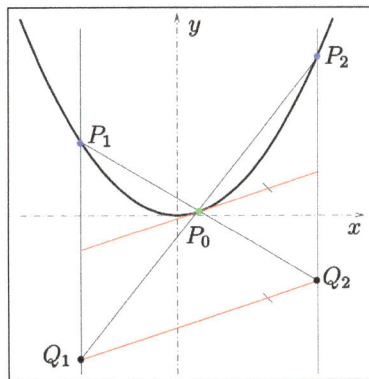

3-points-1-tangent-property.

- Let $P_0 = (x_0, y_0), P_1 = (x_1, y_1), P_2 = (x_2, y_2)$ be three points of the parabola with equation $y = ax^2$ and Q_2 the intersection of the secant line $P_0 P_1$ with the line $x = x_2$ and Q_1 the intersection of the secant line $P_0 P_2$ with the line $x = x_1$, then the tangent at point P_0 is parallel to the line $Q_1 Q_2$.

(The lines $x = x_1$ and $x = x_2$ are parallel to the axis of the parabola.)

Proof: can be performed for the unit parabola $y = x^2$. A short calculation shows: line $Q_1 Q_2$ has slope $2x_0$ which is the slope of the tangent at point P_0.

Application: The 3-points-1-tangent-property of a parabola can be used for the construction of the tangent at point P_0, while P_1, P_2, P_0 are given.

The 3-points-1-tangent-property of a parabola is an affine version of the 4-point-degeneration of Pascal's theorem.

2-points-2-tangents-property

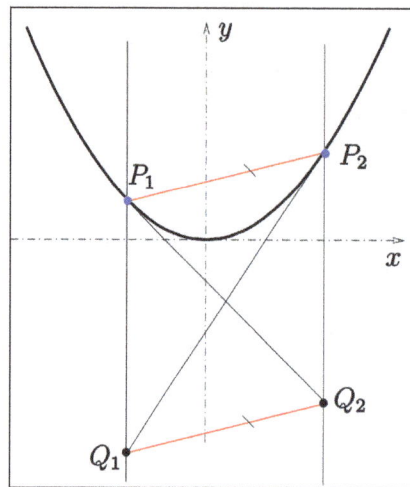

2-points-2-tangents-property.

- Let $P_1 = (x_1, y_1), P_2 = (x_2, y_2)$ be two points of the parabola with equation $y = ax^2$ and Q_2 the intersection of the tangent at point P_1 with the line $x = x_2$ and Q_1 the intersection of the tangent at point P_2 with the line $x = x_1$ then the secant $P_1 P_2$ is parallel to the line $Q_1 Q_2$.

(The lines $x = x_1$ and $x = x_2$ are parallel to the axis of the parabola.)

Proof: straight forward calculation for the unit parabola $y = x^2$.

Application: The 2-points-2-tangents-property can be used for the construction of the tangent of a parabola at point P_2 while P_1, P_2 and the tangent at P_1 are given.

The 2-points-2-tangents-property of a parabola is an affine version of the 3-point-degeneration of Pascal's theorem.

The 2-points-2-tangents-property should not be confused with the following property of a parabola, which deals with 2 points and 2 tangents, too, but is *not* related to Pascal's theorem.

Axis-direction

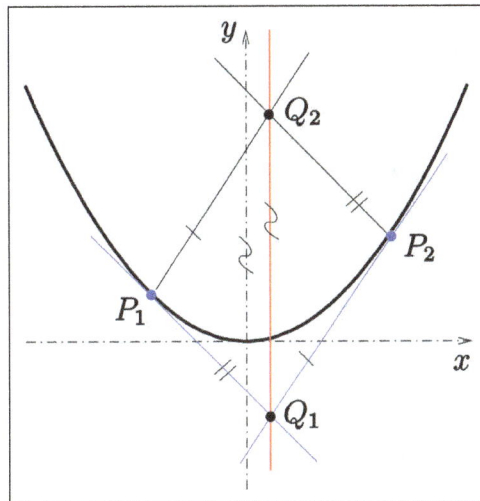

Construction of the axis-direction.

The statements above presume the knowledge of the axis-direction of the parabola, in order to construct the points Q_1, Q_2. The following property determines the points Q_1, Q_2 by two given points and their tangents only, and the result is: the line $Q_1 Q_2$ is parallel to the axis of the parabola.

1. $P_1 = (x_1, y_1), P_2 = (x_2, y_2)$ be two points of the parabola $y = ax^2$ and t_1, t_2 be their tangents;

2. Q_1 be the intersection of the tangents t_1, t_2,

3. Q_2 be the intersection of the parallel line to t_1 through P_2 with the parallel line to t_2 through P_1.

    Then the line $Q_1 Q_2$ is parallel to the axis of the parabola and has the equation $x = \frac{x_1 + x_2}{2}$.

Proof: can be done (like the properties above) for the unit parabola $y = x^2$.

Application: This property can be used to determine the direction of the axis of a parabola, if two points and their tangents are given. An alternative way is to determine the midpoints of two parallel chords.

This property is an affine version of the theorem of two *perspective triangles* of a non-degenerate conic.

Steiner Generation

Parabola

Steiner established the following procedure for the construction of a non-degenerate conic:

Given two pencils $B(U), B(V)$ of lines at two points U, V (all lines containing U and V respectively) and a projective but not perspective mapping π of $B(U)$ onto $B(V)$. Then the intersection points of corresponding lines form a non-degenerate projective conic section.

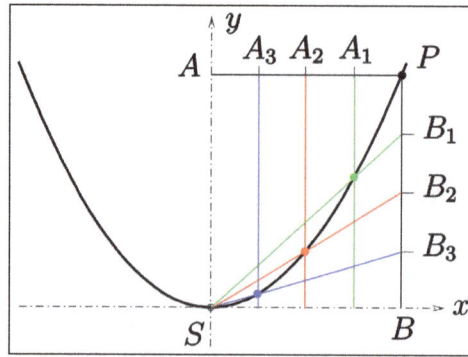

Steiner generation of a parabola.

This procedure can be used for a simple construction of points on the parabola $y = ax^2$:

Consider the pencil at the vertex $S(0,0)$ and the set of lines Π_y, which are parallel to the y-axis.

1. Let $P = (x_0, y_0)$ be a point on the parabola and $A = (0, y_0)$, $B = (x_0, 0)$.

2. The line segment $\overline{BP}$ is divided into n equally spaced segments and this division is project-ed (in the direction BA) onto the line segment $\overline{AP}$. This projection gives rise to a projec-tive mapping S from pencil onto the pencil Π_y.

3. The intersection of the line SB_i and the i-th parallel to the y-axis is a point on the parabola.

Proof: straightforward calculation.

Steiner's generation is also available for ellipses and hyperbolas.

Dual Parabola

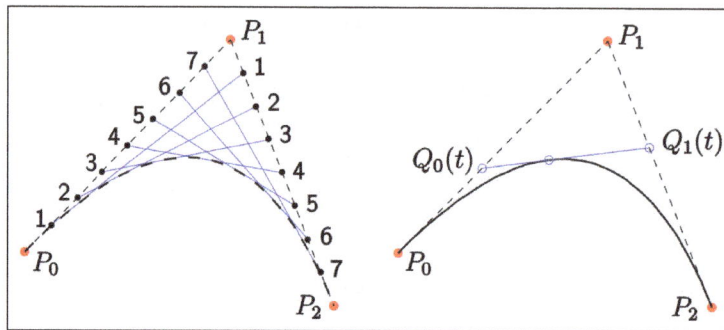

Dual parabola and Bezier curve of degree 2
(right: curve point and division points Q_0, Q_1 for parameter $t = 0.4$)

A *dual parabola* consists of the set of tangents of an ordinary parabola.

The Steiner generation of a conic can be applied to the generation of a dual conic by changing the meanings of points and lines:

- Let be given two point sets on two lines u, v and a projective but not perspective mapping π between these point sets, then the connecting lines of corresponding points form a non degenerate dual conic.

In order to generate elements of a dual parabola, one starts with:

1. Three points P_0, P_1, P_2 not on a line,

2. Divides the line sections $\overline{P_0 P_1}$ and $\overline{P_1 P_2}$ each into n equally spaced line segments and adds numbers as shown in the picture.

3. Then the lines $P_0 P_1, P_1 P_2, (1,1), (2,2), \dots$ are tangents of a parabola, hence elements of a dual parabola.

4. The parabola is a Bezier curve of degree 2 with the control points P_0, P_1, P_2.

The *proof* is a consequence of the *de Casteljau algorithm* for a Bezier curve of degree 2.

Inscribed Angles and the 3-point-form

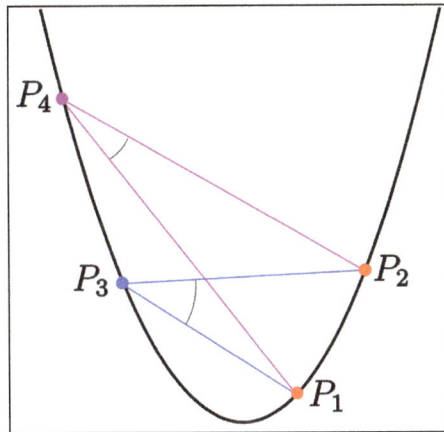

Inscribed angles of a parabola.

A parabola with equation $y = ax^2 + bx + c, a \neq 0$ is uniquely determined by three points $(x_1, y_1), (x_2, y_2), (x_3, y_3)$ with different x-coordinates. The usual procedure to determine the coefficients a, b, c is to insert the point coordinates into the equation. The result is a linear system of three equations, which can be solved by Gaussian elimination or Cramer's rule, for example. An alternative way uses the *inscribed angle theorem* for parabolas:

In the following, the angle of two lines will be measured by the difference of the slopes of the line with respect to the directrix of the parabola. That is, for a parabola of equation $y = ax^2 + bx + c$, the angle between two lines of equations $y = m_1 x + d_1$, $y = m_2 x + d_2$ is measured by $m_1 - m_2$.

Analogous to the inscribed angle theorem for circles one has the *Inscribed angle theorem for parabolas*:

Four points $P_i = (x_i, y_i), i = 1, \dots, 4$, with different x-coordinates, are on a parabola with equation $y = ax^2 + bx + c$ if and only if the angles at P_3 and P_4 have the same measure, as defined above. That is,

$$\frac{y_4 - y_1}{x_4 - x_1} - \frac{y_4 - y_2}{x_4 - x_2} = \frac{y_3 - y_1}{x_3 - x_1} - \frac{y_3 - y_2}{x_3 - x_2}.$$

(Proof: straightforward calculation: If the points are on a parabola, one may translate the coordinates for having the equation $y = ax^2$; then one has $\dfrac{y_i - y_j}{x_i - x_j} = x_i + x_j$ if the points are on the parabola.)

A consequence is that the equation (in x, y) of the parabola determined by 3 points $P_i = (x_i, y_i), i = 1, 2, 3$, with different x-coordinates is (if two x-coordinates are equal there is no parabola with directrix parallel to the x-axis, which passes through the points):

$$\frac{y - y_1}{x - x_1} - \frac{y - y_2}{x - x_2} = \frac{y_3 - y_1}{x_3 - x_1} - \frac{y_3 - y_2}{x_3 - x_2}.$$

Multiplying by the denominators that depend on x, one obtains the more standard form:

$$(x_1 - x_2)y = (x - x_1)(x - x_2)\left(\frac{y_3 - y_1}{x_3 - x_1} - \frac{y_3 - y_2}{x_3 - x_2}\right) + (y_1 - y_2)x + x_1 y_2 - x_2 y_1.$$

Pole-polar Relation

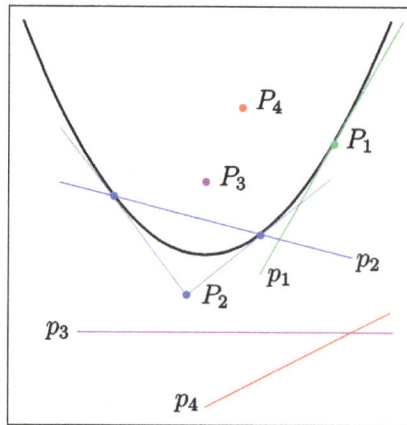

Parabola: pol-polar-relation.

In a suitable coordinate system any parabola can be described by an equation $y = ax^2$. The equation of the tangent at a point $P_0 = (x_0, y_0), y_0 = ax_0^2$ is:

$$y = 2ax_0(x - x_0) + y_0 = 2ax_0 x - ax_0^2 = 2ax_0 x - y_0.$$

One obtains the function:

$$(x_0, y_0) \rightarrow y = 2ax_0 x - y_0$$

on the set of points of the parabola onto the set of tangents.

Obviously this function can be extended onto the set of all points of $\mathbb{R}^2$ to a bijection between the points of $\mathbb{R}^2$ and the lines with equations $y = mx + d, m, d \in \mathbb{R}$. The inverse mapping is:

line $y = mx + d \rightarrow$ point $\left(\frac{m}{2a}, -d\right)$.

This relation is called the *pole-polar relation of the parabola*, where the point is the *pole* and the corresponding line its *polar*.

By calculation one checks the following properties of the pole-polar relation of the parabola:

- For a point (pole) *on* the parabola the polar is the tangent at this point.

- For a pole *P outside* the parabola the intersection points of its polar with the parabola are the touching points of the two tangents passing.

- For a point *within* the parabola the polar has no point with the parabola in common.

- The intersection point of two polar lines (for example: p_3, p_4) is the pole of the connecting line of their poles (in example: P_3, P_4).

- Focus and directrix of the parabola are a pole-polar pair.

Pole-polar relations exist for ellipses and hyperbolas, too.

Tangent Properties

Two Tangent Properties Related to the Latus Rectum

Let the line of symmetry intersect the parabola at point Q, and denote the focus as point F and its distance from point Q as f. Let the perpendicular to the line of symmetry, through the focus, intersect the parabola at a point T. Then (1) the distance from F to T is $2f$, and (2) a tangent to the parabola at point T intersects the line of symmetry at a 45° angle.

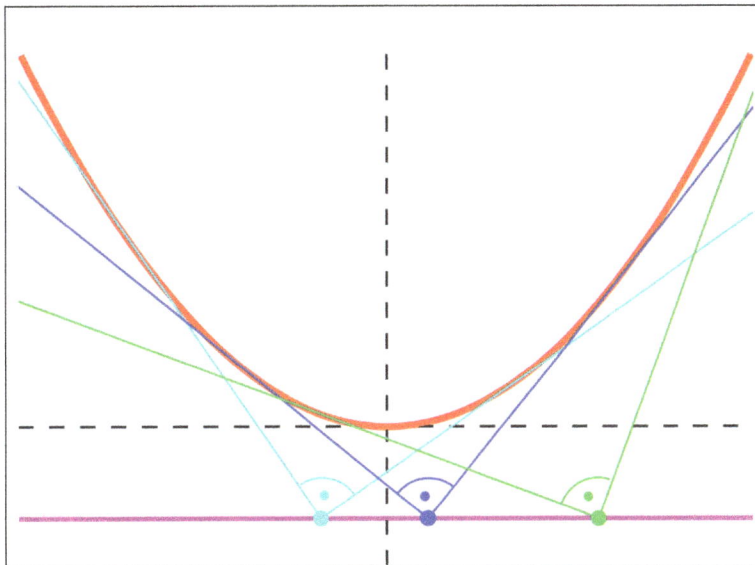

Perpendicular tangents intersect on the directrix.

Orthoptic Property

If two tangents to a parabola are perpendicular to each other, then they intersect on the directrix. Conversely, two tangents which intersect on the directrix are perpendicular.

Lambert's Theorem

Let three tangents to a parabola form a triangle. Then Lambert's theorem states that the focus of the parabola lies on the circumcircle of the triangle.

Tsukerman's converse to Lambert's theorem states that, given three lines that bound a triangle, if two of the lines are tangent to a parabola whose focus lies on the circumcircle of the triangle, then the third line is also tangent to the parabola.

Facts Related to Chords

Focal Length Calculated from Parameters of a Chord

Suppose a chord crosses a parabola perpendicular to its axis of symmetry. Let the length of the chord between the points where it intersects the parabola be c and the distance from the vertex of the parabola to the chord, measured along the axis of symmetry, be d. The focal length, f, of the parabola is given by:

$$f = \frac{c^2}{16d}$$

Suppose a system of Cartesian coordinates is used such that the vertex of the parabola is at the origin, and the axis of symmetry is the y-axis. The parabola opens upward. It is shown elsewhere in this article that the equation of the parabola is $4fy = x^2$, where f is the focal length. At the positive x end of the chord, $x = \left(\frac{c}{2}\right)$ and $y = d$. Since this point is on the parabola, these coordinates must satisfy the equation above. Therefore, by substitution, $4fd = \left(\frac{c}{2}\right)$.

From this, $f = \frac{c^2}{16d}$.

Area Enclosed between a Parabola and a Chord

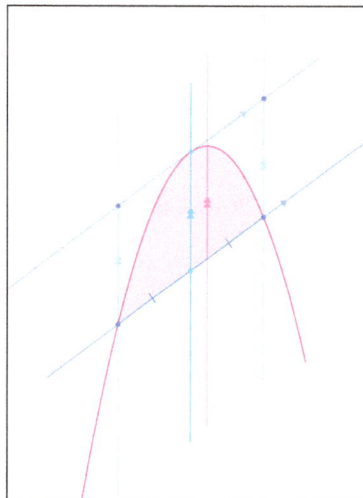

Parabola (magenta) and line (lower light blue) including a chord (blue). The area enclosed between them is in pink. The chord itself ends at the points where the line intersects the parabola.

The area enclosed between a parabola and a chord is two-thirds of the area of a parallelogram which surrounds it. One side of the parallelogram is the chord, and the opposite side is a tangent to the parabola. The slope of the other parallel sides is irrelevant to the area. Often, as here, they are drawn parallel with the parabola's axis of symmetry, but this is arbitrary.

A theorem equivalent to this one, but different in details, was derived by Archimedes in the 3rd Century BCE. He used the areas of triangles, rather than that of the parallelogram.

If the chord has length b, and is perpendicular to the parabola's axis of symmetry, and if the perpendicular distance from the parabola's vertex to the chord is h, the parallelogram is a rectangle, with sides of b and h. The area, A, of the parabolic segment enclosed by the parabola and the chord is therefore:

$$A = \frac{2}{3}bh$$

This formula can be compared with the area of a triangle: $\frac{1}{2}bh$.

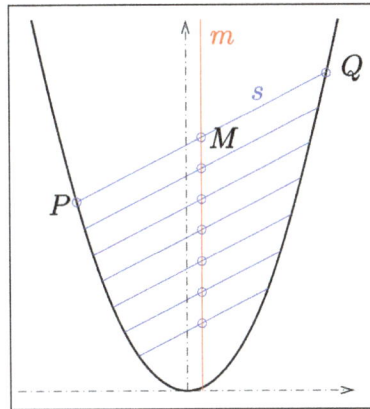
Midpoints of parallel chords.

Corollary Concerning Midpoints and Endpoints of Chords

A corollary of the above discussion is that if a parabola has several parallel chords, their midpoints all lie on a line which is parallel to the axis of symmetry. If tangents to the parabola are drawn through the endpoints of any of these chords, the two tangents intersect on this same line parallel to the axis of symmetry.

Arc Length

If a point X is located on a parabola which has focal length f, and if p is the perpendicular distance from X to the axis of symmetry of the parabola, then the lengths of arcs of the parabola which terminate at X can be calculated from f and p as follows, assuming they are all expressed in the same units.

$$h = \frac{p}{2}$$

$$q = \sqrt{f^2 + h^2}$$

$$s = \frac{hq}{f} + f \ln\left(\frac{h+q}{f}\right)$$

This quantity, s, is the length of the arc between X and the vertex of the parabola.

The length of the arc between X and the symmetrically opposite point on the other side of the parabola is $2s$.

The perpendicular distance, p, can be given a positive or negative sign to indicate on which side of the axis of symmetry X is situated. Reversing the sign of p reverses the signs of h and s without changing their absolute values. If these quantities are signed, the length of the arc between *any* two points on the parabola is always shown by the difference between their values of s. The calculation can be simplified by using the properties of logarithms:

$$s_1 - s_2 = \frac{h_1 q_1 - h_2 q_2}{f} + f \ln \frac{h_1 + q_1}{h_2 + q_2}$$

This can be useful, for example, in calculating the size of the material needed to make a parabolic reflector or parabolic trough.

This calculation can be used for a parabola in any orientation. It is not restricted to the situation where the axis of symmetry is parallel to the y-axis.

A Geometrical Construction to find a Sector Area

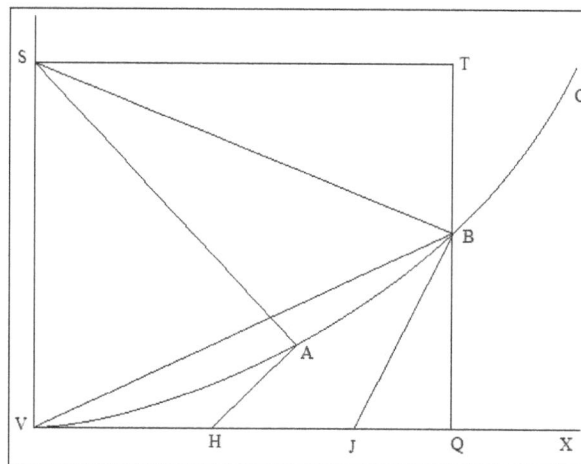

S is the Focus and V is the Principal Vertex of the parabola VG. Draw VX perpendicular to SV.

Take any point B on VG and drop a perpendicular BQ from B to VX. Draw perpendicular ST intersecting BQ, extended if necessary, at T. At B draw the perpendicular BJ, intersecting VX at J.

For the parabola, the segment VBV, the area enclosed by the chord VB and the arc VB, is equal to $\triangle$VBQ / 3, also $BQ = \dfrac{VQ^2}{4SV}$

The Area of the Parabolic Sector SVB = ΔSVB + ΔVBQ / 3:

$$= \frac{SV \cdot VQ}{2} + \frac{VQ \cdot BQ}{6}$$

Since triangles TSB and QBJ are similar:

$$VJ = VQ - JQ = VQ - \frac{BQ \cdot TB}{ST} = VQ - \frac{BQ \cdot (SV - BQ)}{VQ} = \frac{3VQ}{4} + \frac{VQ \cdot BQ}{4SV}$$

Therefore, the Area of the Parabolic Sector $SVB = \frac{2SV \cdot VJ}{3}$, and can be found from the length of VJ, as found above.

A circle through S, V and B also passes through J.

Conversely, if a point, B on the parabola VG is to be found so that the Area of the Sector SVB is equal to a specified value, determine the point J on VX, and construct a circle through S, V and J. Since SJ is the diameter, the center of the circle is at its midpoint, and it lies on the perpendicular bisector of SV, a distance of one half VJ from SV. The point required, B is where this circle intersects the parabola.

If a body traces the path of the parabola due to an inverse square force directed towards S, the area SVB increases at a constant rate as point B moves forward. It follows that J moves at constant speed along VX as B moves along the parabola.

If the speed of the body at the vertex, where it is moving perpendicularly to SV is v, then the speed of J is equal to 3v/4.

The construction can be extended simply to include the case where neither radius coincides with the axis, SV as follows. Let A be a fixed point on VG between V and B, and point H be the intersection on VX with the perpendicular to SA at A. From the above, the Area of the Parabolic Sector

$$SAB = \frac{2SV \cdot (VJ - VH)}{3} = \frac{2SV \cdot HJ}{3}$$

Conversely, if it is required to find the point B for a particular area SAB, find point J from HJ and point B as before. By Book 1 Proposition 16, Corollary 6 of the Principia, the speed of a body moving along a parabola with a force directed towards the focus is inversely proportional to the square root of the radius. If the speed at A is v, then at the vertex, V it is $\sqrt{\frac{SA}{SV}}v$, and point J moves at a constant speed of $\frac{3v}{4}\sqrt{\frac{SA}{SV}}$.

The above construction was devised by Isaac Newton and can be found in Book 1 of the Principia as Proposition 30.

Focal Length and Radius of Curvature at the Vertex

The focal length of a parabola is half of its radius of curvature at its vertex.

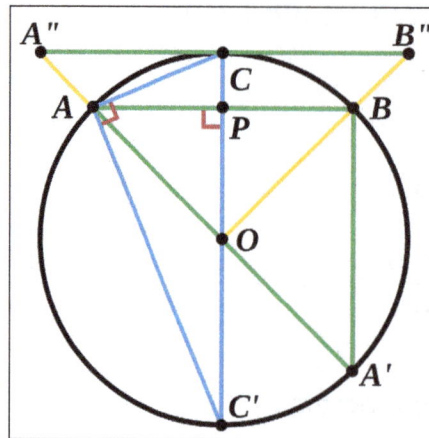

Image is inverted. AB is *x*-axis. C is origin. O is center. A is (x, y).
OA = OC = R. PA = *x*. CP = *y*. OP = $(R - y)$. Other points and lines are irrelevant for this purpose.

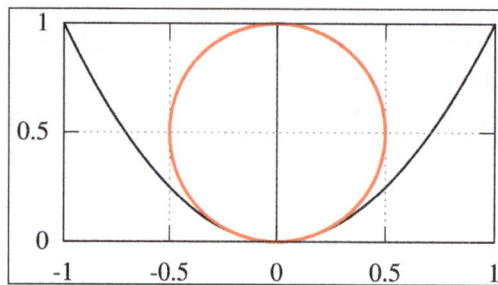

The radius of curvature at the vertex is twice the focal length. The measurements shown on the above diagram are in units of the latus rectum, which is four times the focal length.

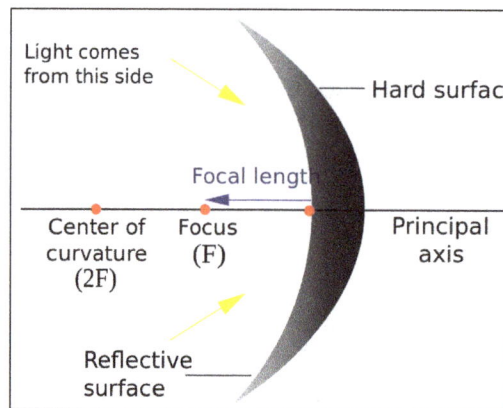

Consider a point (x, y) on a circle of radius R and with center at the point $(0, R)$. The circle passes through the origin. If the point is near the origin, the Pythagorean theorem shows that:

$$x^2 + (R - y)^2 = R^2$$

$$\therefore x^2 + R^2 - 2Ry + y^2 = R^2$$

$$\therefore x^2 + y^2 = 2Ry$$

But if (x, y) is extremely close to the origin, since the *x*-axis is a tangent to the circle, y is very small compared with x, so y^2 is negligible compared with the other terms. Therefore, extremely close to the origin:

$$x \qquad Ry$$

Compare this with the parabola:

$$x^2 = 2Ry$$

which has its vertex at the origin, opens upward, and has focal length f.

Equations 1 and 2 are equivalent if $R = 2f$. Therefore, this is the condition for the circle and parabola to coincide at and extremely close to the origin. The radius of curvature at the origin, which is the vertex of the parabola, is twice the focal length.

Corollary

A concave mirror which is a small segment of a sphere behaves approximately like a parabolic mirror, focusing parallel light to a point which is midway between the centre and the surface of the sphere.

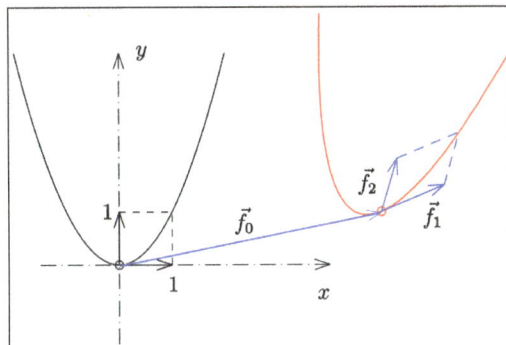

Parabola as an affine image of the unit parabola.

Another definition of a parabola uses affine transformations:

- Any *parabola* is the affine image of the unit parabola with equation $y = x^2$.

An affine transformation of the Euclidean plane has the form $\vec{x} \rightarrow \vec{f}_0 + A\vec{x}$, where A is a regular matrix (determinant is not 0) and $\vec{f}_0$ is an arbitrary vector. If $\vec{f}_1, \vec{f}_2$ are the column vectors of the matrix A, the unit parabola $(t, t^2), t \in \mathbb{R}$, is mapped onto the parabola:

$$\vec{x} = \vec{p}(t) = \vec{f}_0 + \vec{f}_1 t + \vec{f}_2 t^2$$

where,

$\vec{f}_0$ is a *point* of the parabola and,

$\vec{f}_1$ is a *tangent vector* at point,

$\vec{f}_2$ is *parallel to the axis* of the parabola (axis of symmetry through the vertex).

In general the two vectors $\vec{f}_1, \vec{f}_2$ are not perpendicular and $\vec{f}_0$ is *not* the vertex, unless the affine transformation is a similarity.

The tangent vector at the point $\vec{p}(t)$ is $\vec{p}'(t) = \vec{f}_1 + 2t\vec{f}_2$. At the vertex the tangent vector is orthogonal to $\vec{f}_2$. Hence the parameter t_0 of the vertex is the solution of the equation $\vec{p}'(t) \cdot \vec{f}_2 = \vec{f}_1 \cdot \vec{f}_2 + 2t\vec{f}_2^2 = 0$, which is $t_0 = -\frac{\vec{f}_1 \cdot \vec{f}_2}{2\vec{f}_2^2}$ and

$$\vec{p}(t_0) = \vec{f}_0 - \frac{\vec{f}_1 \cdot \vec{f}_2}{2\vec{f}_2^2} \vec{f}_1 + \frac{(\vec{f}_1 \cdot \vec{f}_2)^2}{4(\vec{f}_2^2)^2} \vec{f}_2 \text{ is the vertex.}$$

The *focal length* can be determined by a suitable parameter transformation (which does not change the geometric shape of the parabola). The focal length is

$$f = \frac{\vec{f}_1^2 \cdot \vec{f}_2^2 - (\vec{f}_1 \cdot \vec{f}_2)^2}{4|\vec{f}_2|^3} .$$

Hence

$$F : \vec{f}_0 - \frac{\vec{f}_1 \cdot \vec{f}_2}{2\vec{f}_2^2} \vec{f}_1 + \frac{\vec{f}_1^2 \cdot \vec{f}_2^2}{4(\vec{f}_2^2)^2} \vec{f}_2 \text{ is the } \textit{focus} \text{ of the parabola.}$$

Remark: The advantage of this definition is, one obtains a simple parametric representation of an arbitrary parabola, even in the space, if the vectors $\vec{f}_0, \vec{f}_1, \vec{f}_2$ are vectors of the Euclidean space.

As Quadratic Bézier Curve

A quadratic Bézier curve is a curve $\vec{c}(t)$ defined by three points $P_0 : \vec{p}_0$, $P_1 : \vec{p}_1$ and $P_2 : \vec{p}_2$, its *control points*:

$$\vec{c}(t) = \sum_{i=0}^{2} \binom{2}{i} t^i (1-t)^{2-i} \vec{p}_i$$

$$= (1-t)^2 \vec{p}_0 + 2t(1-t)\vec{p}_1 + t^2 \vec{p}_2$$

$$= (\vec{p}_0 - 2\vec{p}_1 + \vec{p}_2)t^2 + (-2\vec{p}_0 + 2\vec{p}_1)t + \vec{p}_0 , \; t \in [0,1]$$

This curve is an arc of a parabola

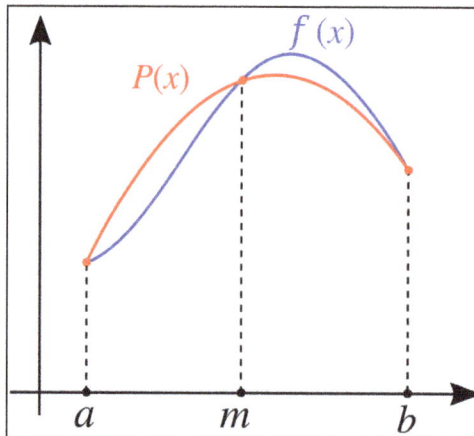

Simpson's rule: the graph of a function is replaced by an arc of a parabola.

Numerical Integration

In one method of numerical integration one replaces the graph of a function by arcs of parabolas and integrates the parabola arcs. A parabola is determined by three points. The formula for one arc is:

$$\int_a^b f(x)dx \approx \frac{b-a}{6} \cdot \left(f(a) + 4f\left(\frac{a+b}{2}\right) + f(b) \right).$$

The method is called Simpson's rule.

As Trisectrix

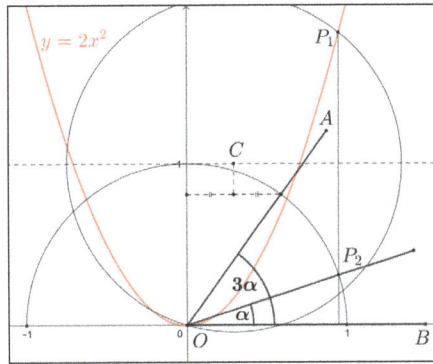

Angle trisection with a parabola.

A parabola can be used as a trisectrix, that is it allows the exact trisection of an arbitrary angle with straightedge and compass. Note that this is not in contradiction to the impossibility of an angle trisection with compass-and-straightedge constructions alone, as the use of parabolas is not allowed in the classic rules for compass-and-straightedge constructions.

To trisect $\angle AOB$ place its leg OB on the x-axis such that the vertex O is in the coordinate system's origin. The coordinate system also contains the parabola $y = 2x^2$. The unit circle with radius 1 around the origin intersects the angle's other leg OA and from this point of intersection draw the perpendicular onto the y-axis. The parallel to y-axis through the midpoint of that perpendicular and the tangent on the unit circle in $(0,1)$ intersect in C. The circle around C with radius OC intersects the parabola in P_1. The perpendicular from P_1 onto the x-axis intersects the unit circle in P_2 and $\angle P_2 OB$ is exactly one third of $\angle AOB$.

The correctness of this construction can be seen by showing that the x-coordinate of P_1 is $\cos(\alpha)$. Solving the equation system given by the circle around C and the parabola leads to the cubic equation $4x^3 - 3x - \cos(3\alpha) = 0$. The triple angle formula $\cos(3\alpha) = 4\cos(\alpha)^3 - 3\cos(\alpha)$ then shows that $\cos(\alpha)$ is indeed a solution of that cubic equation.

CIRCLE

Circle is one of the conic sections, consisting of the set of all points the same distance (the radius) from a given point (the centre). A line connecting any two points on a circle is called a chord, and a chord passing through the centre is called a diameter. The distance around a circle (the circumference) equals the length of a diameter multiplied by π. The area of a circle is the square of the radius multiplied by π. An arc consists of any part of a circle encompassed by an angle with its vertex at the centre (central angle). Its length is in the same proportion to the circumference as the central angle is to a full revolution.

Important Terms Related to Circle

Here are a few important terms that make up for the parts of a circle.

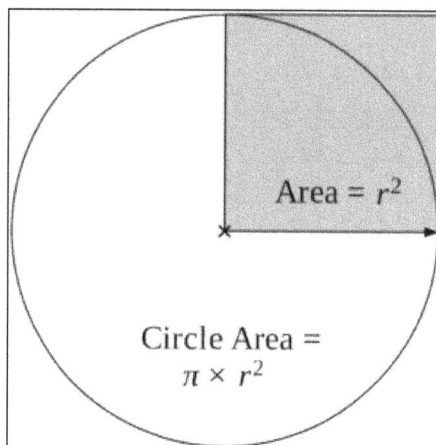

Area = r^2

Circle Area =
$\pi \times r^2$

- Diameter: The diameter is a line which is drawn across a circle passing through the centre.

- Radius: The distance from the middle or centre of a circle towards any point on it is a radius. Interestingly, when you place two radii back-to-back, the resultant would hold the same length as one diameter. Therefore, we can call one diameter twice as long as the concerned radius.

- Area of Circle: In a circle, the area can be stated as π times the square of the radius. It is, $A = \pi r^2$. Taking into consideration the diameter: $A = (\pi/4) \times D^2$

- Chord: A line segment that joins two points present on a curve is the chord. In geometry, the usefulness of a chord is focused on describing a line segment connecting two endpoints which rest on a circle.

- Tangent & Arc: A line which slightly touches the circle on its travel to a different direction is Tangent. On the other hand, a part of the circumference is an Arc.

- Sector & Segment: A sector is a part of a circle surrounded by two radii of it together with their intercepted arc. The segment is that region which is enclosed by a chord together with the arc subtended by the chord.

Properties of the Circle

- Circles with equal radii are congruent.

- Also, the circles with different radii appear to be similar.

- The chords that are equidistant from the centre are of the same length.

- All points on the circle are equidistant from the centre point.

- The longest chord in the circle is the diameter.

- A diameter of a circle divides it into two equal arcs. Each of the arcs is s a semi-circle.

- If the radii of two circles are exactly the same value, then the circles are congruent.

- Two or more circles that have different radii but the same centre are concentric circles.

ELLIPSE

In mathematics, an ellipse is a plane curve surrounding two focal points, such that for all points on the curve, the sum of the two distances to the focal points is a constant. As such, it generalizes a circle, which is the special type of ellipse in which the two focal points are the same. The elongation of an ellipse is measured by its eccentricity e, a number ranging from $e = 0$ (the limiting case of a circle) to $e = 1$ (the limiting case of infinite elongation, no longer an ellipse but a parabola).

Analytically, the equation of a standard ellipse centered at the origin with width $2a$ and height $2b$ is:

$$\frac{x^2}{a^2} + \frac{y^2}{b^2} = 1.$$

Assuming $a \geq b$, the foci are $(\pm c, 0)$ for $c = \sqrt{a^2 - b^2}$. The standard parametric equation is:

$$(x, y) = (a\cos(t), b\sin(t)) \quad \text{for} \quad 0 \leq t \leq 2\pi.$$

Ellipses are the closed type of conic section: a plane curve tracing the intersection of a cone with a plane. Ellipses have many similarities with the other two forms of conic sections, parabolas and hyperbolas, both of which are open and unbounded. An angled cross section of a cylinder is also an ellipse.

An ellipse may also be defined in terms of one focus point and a line outside the ellipse called the directrix: for all points on the ellipse, the ratio between the distance to the focus and the distance to the directrix is a constant. This constant ratio is the above-mentioned eccentricity $e = \frac{c}{a} = \sqrt{1 - \frac{b^2}{a^2}}$.

Ellipses are common in physics, astronomy and engineering. For example, the orbit of each planet in the solar system is approximately an ellipse with the Sun at one focus point (more precisely, the focus is the barycenter of the Sun–planet pair). The same is true for moons orbiting planets and all other systems of two astronomical bodies. The shapes of planets and stars are often well described by ellipsoids. A circle viewed from a side angle looks like an ellipse: that is, the ellipse is the image of a circle under parallel or perspective projection. The ellipse is also the simplest Lissajous figure formed when the horizontal and vertical motions are sinusoids with the same frequency: a similar effect leads to elliptical polarization of light in optics.

Locus of Points

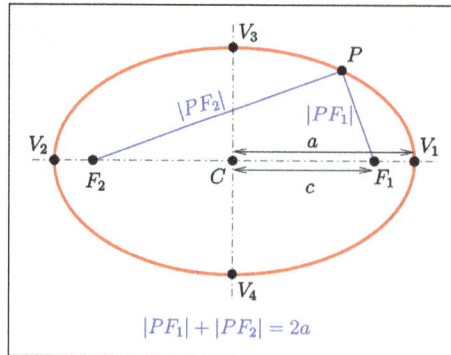

Ellipse: Definition by sum of distances to foci.

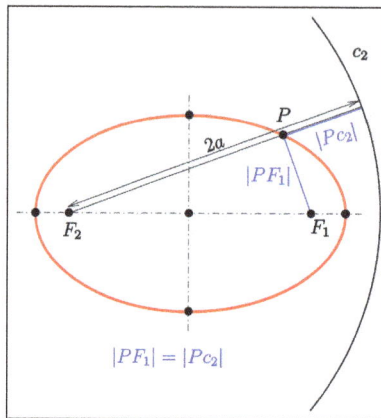

Ellipse: Definition by focus and circular directrix.

An ellipse can be defined geometrically as a set or locus of points in the Euclidean plane:

- Given two fixed points F_1, F_2 called the foci and a distance $2a$ which is greater than the distance between the foci, the ellipse is the set of points P such that the sum of the distances $|PF_1|, |PF_2|$ is equal to $2a$: $E = \{P \in \mathbb{R}^2 \check{Z} PF_2| + |PF_1| = 2a\}$.

The midpoint C of the line segment joining the foci is called the *center* of the ellipse. The line through the foci is called the *major axis*, and the line perpendicular to it through the center is the *minor axis*. The major axis intersects the ellipse at the *vertex* points V_1, V_2, which have distance a to the center. The distance c of the foci to the center is called the *focal distance* or linear eccentricity. The quotient $e = \frac{c}{a}$ is the *eccentricity*.

The case $F_1 = F_2$ yields a circle and is included as a special type of ellipse.

The equation $|PF_2| + |PF_1| = 2a$ can be viewed in a different way: If c_2 is the circle with midpoint F_2 and radius $2a$, then the distance of a point P to the circle c_2 equals the distance to the focus F_1:

$$|PF_1| = |Pc_2|.$$

c_2 is called the *circular directrix* (related to focus F_2) of the ellipse. This property should not be confused with the definition of an ellipse using a directrix line below.

Using Dandelin spheres, one can prove that any plane section of a cone with a plane is an ellipse, assuming the plane does not contain the apex and has slope less than that of the lines on the cone.

In Cartesian Coordinates

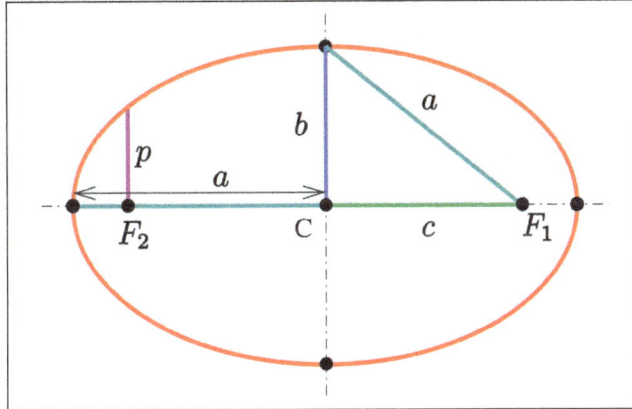

Shape parameters: a:semi-major axis, b: semi-minor axis
c: linear eccentricity, p: semi-latus rectum (usually ℓ).

Standard Equation

The standard form of an ellipse in Cartesian coordinates assumes that the origin is the center of the ellipse, the x-axis is the major axis, and:

- The foci are the points $F_1 = (c,0), F_2 = (-c,0)$,

- The vertices are $V_1 = (a,0), V_2 = (-a,0)$.

For an arbitrary point (x, y) the distance to the focus $(c,0)$ is $\sqrt{(x-c)^2 + y^2}$ and to the other focus $\sqrt{(x+c)^2 + y^2}$. Hence the point (x, y) is on the ellipse whenever:

$$\sqrt{(x-c)^2 + y^2} + \sqrt{(x+c)^2 + y^2} = 2a.$$

Removing the radicals by suitable squarings and using $b^2 = a^2 - c^2$ produces the standard equation of the ellipse:

$$\frac{x^2}{a^2} + \frac{y^2}{b^2} = 1,$$

or, solved for y:

$$y = \pm \frac{b}{a}\sqrt{a^2 - x^2} = \pm \sqrt{(a^2 - x^2)(1 - e^2)}.$$

The width and height parameters a, b are called the semi-major and semi-minor axes. The top and bottom points $V_3 = (0,b), V_4 = (0,-b)$ are the *co-vertices*. The distances from a point (x, y) on the ellipse to the left and right foci are $a + ex$ and $a - ex$.

It follows from the equation that the ellipse is *symmetric* with respect to the coordinate axes and hence with respect to the origin.

Parameters

Semi-major and Semi-minor Axes $a \geq b$

Throughout this article a is the semi-major axis, i.e. $a \geq b > 0$. In general the canonical ellipse equation $\frac{x^2}{a^2} + \frac{y^2}{b^2} = 1$ may have $a < b$ (and hence the ellipse would be taller than it is wide); in this form the semi-major axis would be b. This form can be converted to the standard form by transposing the variable names x and y and the parameter names a and b.

Linear Eccentricity c

This is the distance from the center to a focus: $c = \sqrt{a^2 - b^2}$.

Eccentricity e

The eccentricity can be expressed as:

$$e = \frac{c}{a} = \sqrt{1 - \left(\frac{b}{a}\right)^2},$$

assuming $a > b$. An ellipse with equal axes ($a = b$) has zero eccentricity, and is a circle.

Semi-latus Rectum l

The length of the chord through one of the foci, perpendicular to the major axis, is called the *latus rectum*. One half of it is the *semi-latus rectum* ℓ. A calculation shows:

$$\ell = \frac{b^2}{a} = a(1 - e^2).$$

The semi-latus rectum ℓ is equal to the *radius of curvature* of the osculating circles at the vertices.

Tangent

An arbitrary line g intersects an ellipse at 0, 1, or 2 points, respectively called an *exterior line*, *tangent* and *secant*. Through any point of an ellipse there is a unique tangent. The tangent at a point (x_1, y_1) of the ellipse $\frac{x^2}{a^2} + \frac{y^2}{b^2} = 1$ has the coordinate equation:

$$\frac{x_1}{a^2} x + \frac{y_1}{b^2} y = 1.$$

A vector parametric equation of the tangent is:

$$\vec{x} = \begin{pmatrix} x_1 \\ y_1 \end{pmatrix} + s \begin{pmatrix} -y_1 a^2 \\ x_1 b^2 \end{pmatrix} \text{ with } s \in \mathbb{R}.$$

Proof: Let be (x_1, y_1) an ellipse point and $\vec{x} = \begin{pmatrix} x_1 \\ y_1 \end{pmatrix} + s \begin{pmatrix} u \\ v \end{pmatrix}$ the equation of any line g containing

(x_1, y_1). Inserting the line's equation into the ellipse equation and respecting $\frac{x_1^2}{a^2} + \frac{y_1^2}{b^2} = 1$ yields:

$$\frac{(x_1 + su)^2}{a^2} + \frac{(y_1 + sv)^2}{b^2} = 1 \quad \Rightarrow \quad 2s\left(\frac{x_1 u}{a^2} + \frac{y_1 v}{b^2}\right) + s^2\left(\frac{u^2}{a^2} + \frac{v^2}{b^2}\right) = 0.$$

- $\frac{x_1}{a^2}u + \frac{y_1}{b^2}v = 0$. Then line g and the ellipse have only point (x_1, y_1) in common, and g is a tangent. The tangent direction has perpendicular vector $(\frac{x_1}{a^2}, \frac{y_1}{b^2})$, so the tangent line has equation $\frac{x_1}{a^2}x + \frac{y_1}{b^2}y = k$ for some k. Because (x_1, y_1) is on the tangent and the ellipse, one obtains $k = 1$.

- $\frac{x_1}{a^2}u + \frac{y_1}{b^2}v \neq 0$. Then line g has a second point in common with the ellipse, and is a secant.

Using just above equation one finds that $(-y_1 a^2, x_1 b^2)$ is a tangent vector at point (x_1, y_1), which proves the vector equation.

If (x_1, y_1) and (u, v) are two points of the ellipse such that $\frac{x_1 u}{a^2} + \frac{y_1 v}{b^2} = 0$, then the points lie on two *conjugate diameters*. (If $a = b$, the ellipse is a circle and "conjugate" means "orthogonal".)

Shifted Ellipse

If the standard ellipse is shifted to have center $(x_\circ, y_\circ)$, its equation is:

$$\frac{(x - x_\circ)^2}{a^2} + \frac{(y - y_\circ)^2}{b^2} = 1.$$

The axes are still parallel to the x- and y-axes.

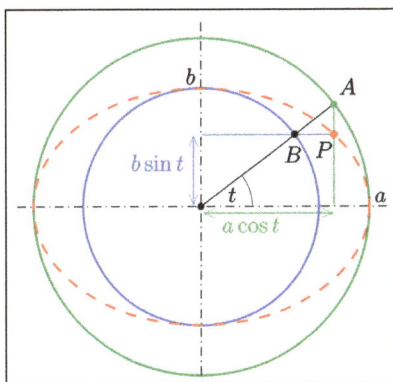

The construction of points based on the parametric equation and the interpretation of parameter t, which is due to de la Hire.

General Ellipse

In analytic geometry, the ellipse is defined as a quadric: the set of points (X, Y) of the Cartesian plane that, in non-degenerate cases, satisfy the implicit equation.

$$AX^2 + BXY + CY^2 + DX + EY + F = 0$$

provided $B^2 - 4AC < 0$.

To distinguish the degenerate cases from the non-degenerate case, let Δ be the determinant

$$\Delta = \begin{vmatrix} A & B/2 & D/2 \\ B/2 & C & E/2 \\ D/2 & E/2 & F \end{vmatrix} = \left(AC - \frac{B^2}{4} \right) F + \frac{BED}{4} - \frac{CD^2}{4} - \frac{AE^2}{4}.$$

Then the ellipse is a non-degenerate real ellipse if and only if $C\Delta < 0$. If $C\Delta > 0$, we have an imaginary ellipse, and if $\Delta = 0$, we have a point ellipse.

The general equation's coefficients can be obtained from known semi-major axis a, semi-minor axis b, center coordinates (x_0, y_0), and rotation angle Θ (the angle from the positive horizontal axis to the ellipse's major axis) using the formulae:

$$A = a^2 (\sin\Theta)^2 + b^2 (\cos\Theta)^2$$

$$B = 2(b^2 - a^2)\sin\Theta\cos\Theta$$

$$C = a^2 (\cos\Theta)^2 + b^2 (\sin\Theta)^2$$

$$D = -2Ax_0 - By_0$$

$$E = -Bx_0 - 2Cy_0$$

$$F = Ax_0^2 + Bx_0 y_0 + Cy_0^2 - a^2 b^2.$$

These expressions can be derived from the canonical equation $\frac{x^2}{a^2} + \frac{y^2}{b^2} = 1$ by an affine transformation of the coordinates (x, y):

$$y = -(X - x_0)\sin\Theta + (Y - y_0)\cos\Theta.$$

$$y = -(X - x_0)\sin\Theta + (Y - y_0)\cos\Theta.$$

Conversely, the canonical form parameters can be obtained from the general form coefficients by the equations:

$$a, b = \frac{\sqrt{2\left(AE^2 + CD^2 - BDE + (B^2 - 4AC)F \right)\left((A+C) \pm \sqrt{(A-C)^2 + B^2} \right)}}{B^2 - 4AC}$$

$$x_0 = \frac{2CD - BE}{B^2 - 4AC}$$

$$y_\circ = \frac{2AE - BD}{B^2 - 4AC}$$

$$\Theta = \begin{cases} 0 & \text{for } B = 0,\ A < C \\ 90^\circ & \text{for } B = 0,\ A > C \\ \arctan \dfrac{C - A - \sqrt{(A-C)^2 + B^2}}{B} & \text{for } B \neq 0. \end{cases}$$

Parametric Representation

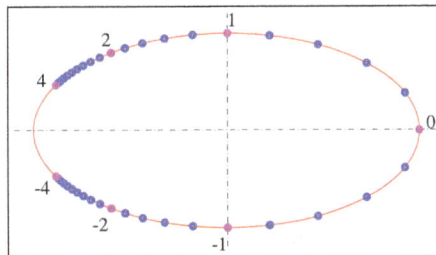

Ellipse points calculated by the rational representation with equal spaced parameters ($\Delta u = 0.2$).

Standard Parametric Representation

Using trigonometric functions, a parametric representation of the standard ellipse $\frac{x^2}{a^2} + \frac{y^2}{b^2} = 1$ is:

$$(x, y) = (a \cos t, b \sin t),\ 0 \leq t < 2\pi.$$

The parameter t (called the *eccentric anomaly* in astronomy) is not the angle of $(x(t), y(t))$ with the x-axis, but has a geometric meaning due to Philippe de La Hire.

Rational Representation

With the substitution $u = \tan(t/2)$ and trigonometric formulae one obtains:

$$\cos t = (1 - u^2)/(u^2 + 1), \quad \sin t = 2u/(u^2 + 1)$$

and the *rational* parametric equation of an ellipse:

$$\begin{aligned} x(u) &= a(1 - u^2)/(u^2 + 1) \\ y(u) &= 2bu/(u^2 + 1) \end{aligned}, \quad -\infty < u < \infty,$$

which covers any point of the ellipse $\frac{x^2}{a^2} + \frac{y^2}{b^2} = 1$ except the left vertex $(-a, 0)$.

For $u \in [0, 1]$, this formula represents the right upper quarter of the ellipse moving counter-clockwise with increasing u. The left vertex is the limit $\lim\limits_{u \to \pm\infty}(x(u), y(u)) = (-a, 0)$.

Rational representations of conic sections are commonly used in Computer Aided Design.

Tangent Slope as Parameter

A parametric representation, which uses the slope m of the tangent at a point of the ellipse can be obtained from the derivative of the standard representation $\vec{x}(t) = (a\cos t, b\sin t)^T$:

$$\vec{x}'(t) = (-a\sin t, b\cos t)^T \quad \rightarrow \quad m = -\frac{b}{a}\cot t \quad \rightarrow \quad \cot t = -\frac{ma}{b}.$$

With help of trigonometric formulae one obtains:

$$\cos t = \frac{\cot t}{\pm\sqrt{1+\cot^2 t}} = \frac{-ma}{\pm\sqrt{m^2 a^2 + b^2}}, \qquad \sin t = \frac{1}{\pm\sqrt{1+\cot^2 t}} = \frac{b}{\pm\sqrt{m^2 a^2 + b^2}}.$$

Replacing $\cos t$ and $\sin t$ of the standard representation yields:

$$\vec{c}_\pm(m) = \left(-\frac{ma^2}{\pm\sqrt{m^2 a^2 + b^2}}, \frac{b^2}{\pm\sqrt{m^2 a^2 + b^2}}\right), m \in \mathbb{R}.$$

Here m is the slope of the tangent at the corresponding ellipse point, $\vec{c}_+$ is the upper and $\vec{c}_-$ the lower half of the ellipse. The vertices $(\pm a, 0)$, having vertical tangents, are not covered by the representation. The equation of the tangent at point $\vec{c}_\pm(m)$ has the form $y = mx + n$. The still unknown can be determined by inserting the coordinates of the corresponding ellipse point $\vec{c}_\pm(m)$:

$$y = mx \pm \sqrt{m^2 a^2 + b^2}.$$

This description of the tangents of an ellipse is an essential tool for the determination of the orthoptic of an ellipse.

General Ellipse

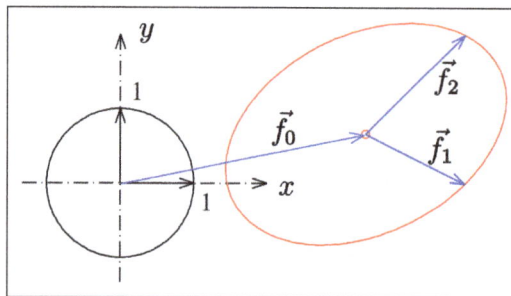

Ellipse as an affine image of the unit circle.

Another definition of an ellipse uses affine transformations:

- Any *ellipse* is an affine image of the unit circle with equation $x^2 + y^2 = 1$.

An affine transformation of the Euclidean plane has the form $\vec{x} \mapsto \vec{f}_0 + A\vec{x}$, where A is a regular matrix (with non-zero determinant) and $\vec{f}_0$ is an arbitrary vector. If $\vec{f}_1, \vec{f}_2$ are the column vectors of the matrix A, the unit circle $(\cos(t), \sin(t))$, $0 \le t \le 2\pi$, is mapped onto the ellipse:

$$\vec{x}=\vec{p}(t)=\vec{f}_0+\vec{f}_1\cos t+\vec{f}_2\sin t.$$

Here $\vec{f}_0$ is the center and $\vec{f}_1,\vec{f}_2$ are the directions of two conjugate diameters, in general not perpendicular. The four vertices of the ellipse are $\vec{p}(t_0),\ \vec{p}(t_0\pm\frac{\pi}{2}),\ \vec{p}(t_0+\pi)$, for a parameter $t=t_0$ defined by:

$$\cot(2t_0)=\frac{\vec{f}_1^{\,2}-\vec{f}_2^{\,2}}{2\vec{f}_1\cdot\vec{f}_2}.$$

(If $\vec{f}_1\cdot\vec{f}_2=0$, then $t_0=0$.) This is derived as follows. The tangent vector at point $\vec{p}(t)$ is:

$$\vec{p}(t)=\vec{f}_1\sin t+\vec{f}_2\cos t.$$

At a vertex parameter $t=t_0$, the tangent is perpendicular to the major/minor axes, so:

$$0=\vec{p}\,'(t)\cdot(\vec{p}(t)-\vec{f}_0)=-(\vec{f}_1\sin t+\vec{f}_2\cos t)\cdot(\vec{f}_1\cos t+\vec{f}_2\sin t).$$

Expanding and applying the identities $\cos^2 t-\sin^2 t=\cos 2t,\ 2\sin t\cos t=\sin 2t$ gives the equation for $t=t_0$.

This definition gives a parametric representation of an arbitrary ellipse, even in space, if we allow $\vec{f}_0,\vec{f}_1,\vec{f}_2$ to be vectors in space.

Polar Forms

Polar Form Relative to Center

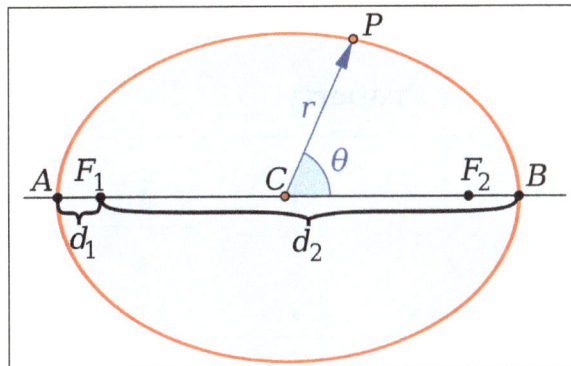

Polar coordinates centered at the center.

In polar coordinates, with the origin at the center of the ellipse and with the angular coordinate θ measured from the major axis, the ellipse's equation is:

$$r(\theta)=\frac{ab}{\sqrt{(b\cos\theta)^2+(a\sin\theta)^2}}=\frac{b}{\sqrt{1-(e\cos\theta)^2}}$$

Polar Form Relative to Focus

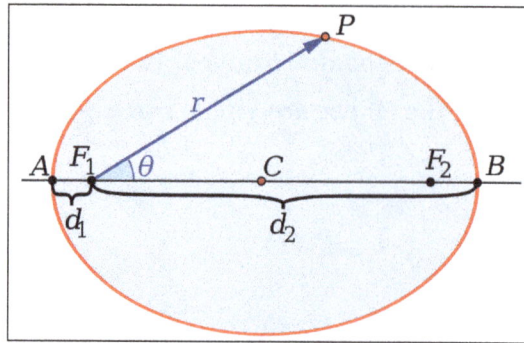

Polar coordinates centered at focus.

If instead we use polar coordinates with the origin at one focus, with the angular coordinate $\theta = 0$ still measured from the major axis, the ellipse's equation is:

$$r(\theta) = \frac{a(1-e^2)}{1 \pm e \cos \theta}$$

where the sign in the denominator is negative if the reference direction $\theta = 0$ points towards the center (as illustrated on the right), and positive if that direction points away from the center.

In the slightly more general case of an ellipse with one focus at the origin and the other focus at angular coordinate ϕ, the polar form is:

$$r = \frac{a(1-e^2)}{1 - e \cos(\theta - \phi)}.$$

The angle θ in these formulas is called the true anomaly of the point. The numerator of these formulas is the semi-latus rectum $\ell = a(1-e^2)$.

Eccentricity and the Directrix Property

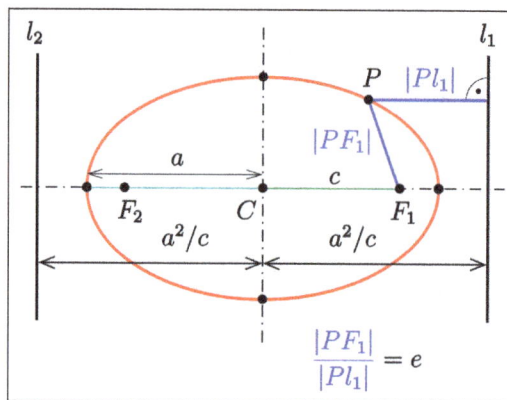

Ellipse: directrix property.

Each of the two lines parallel to the minor axis, and at a distance of $d = \dfrac{a^2}{c} = \dfrac{a}{e}$ from it, is called a *directrix* of the ellipse.

- For an arbitrary point P of the ellipse, the quotient of the distance to one focus and to the corresponding directrix is equal to the eccentricity:

$$\frac{|PF_1|}{|Pl_1|} = \frac{|PF_2|}{|Pl_2|} = e = \frac{c}{a}.$$

The proof for the pair F_1, l_1 follows from the fact that $|PF_1|^2 = (x-c)^2 + y^2, |Pl_1|^2 = \left(x - \frac{a^2}{c}\right)^2$ and $y^2 = b^2 - \frac{b^2}{a^2}x^2$ satisfy the equation:

$$|PF_1|^2 - \frac{c^2}{a^2}|Pl_1|^2 = 0.$$

The second case is proven analogously.

The converse is also true and can be used to define an ellipse (in a manner similar to the definition of a parabola):

- For any point F (focus), any line l (directrix) not through F, and any real number e with $0 < e < 1$, the ellipse is the locus of points for which the quotient of the distances to the point and to the line is e, that is:

$$E = \left\{ P \,\middle|\, \frac{|PF|}{|Pl|} = e \right\}.$$

The choice $e = 0$, which is the eccentricity of a circle, is not allowed in this context. One may consider the directrix of a circle to be the line at infinity.

(The choice $e = 1$ yields a parabola, and if $e > 1$, a hyperbola.)

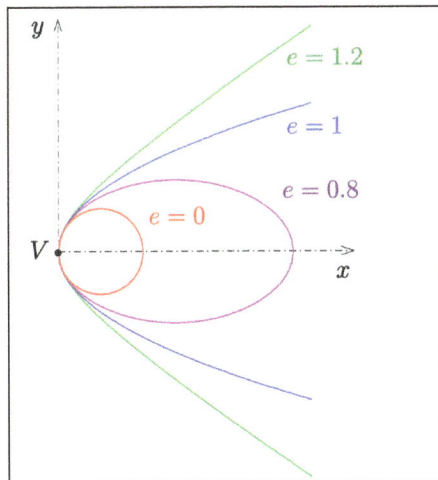

Pencil of conics with a common vertex and common semi-latus rectum.

Let $F = (f, 0), e > 0$, and assume $(0,0)$ is a point on the curve. The directrix l has equation $x = -\frac{f}{e}$. With $P = (x, y)$, the relation $|PF|^2 = e^2 |Pl|^2$ produces the equations

$$(x-f)^2 + y^2 = e^2 \left(x + \frac{f}{e}\right)^2 = (ex + f)^2 \text{ and } x^2(e^2 - 1) + 2xf(1+e) - y^2 = 0.$$

The substitution $p = f(1+e)$ yields

- $x^2(e^2 - 1) + 2px - y^2 = 0.$

This is the equation of an *ellipse* ($e < 1$), or a *parabola* ($e = 1$), or a *hyperbola* ($e > 1$). All of these non-degenerate conics have, in common, the origin as a vertex.

If $e < 1$, introduce new parameters a, b so that $1 - e^2 = \frac{b^2}{a^2}$, and $p = \frac{b^2}{a}$, and then the equation above becomes

$$\frac{(x-a)^2}{a^2} + \frac{y^2}{b^2} = 1,$$

which is the equation of an ellipse with center $(a, 0)$, the x-axis as major axis, and the major/minor semi axis a, b.

If the focus is $F = (f_1, f_2)$ and the directrix $ux + vy + w = 0$, one obtains the equation

$$\left(x - f_1\right)^2 + \left(y - f_2\right)^2 = e^2 \cdot \frac{\left(ux + vy + w\right)^2}{u^2 + v^2}.$$

(The right side of the equation uses the Hesse normal form of a line to calculate the distance $|Pl|$.)

Focus-to-focus Reflection Property

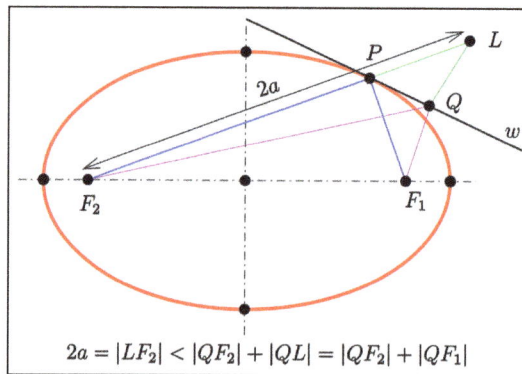

$$2a = |LF_2| < |QF_2| + |QL| = |QF_2| + |QF_1|$$

Ellipse: the tangent bisects the supplementary angle of the angle between the lines to the foci.

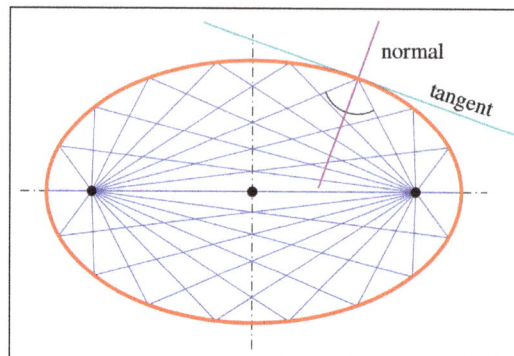

Rays from one focus reflect off the ellipse to pass through the other focus.

An ellipse possesses the following property:

- The normal at a point P bisects the angle between the lines $\overline{PF_1}, \overline{PF_2}$.

Because the tangent is perpendicular to the normal, the statement is true for the tangent and the supplementary angle of the angle between the lines to the foci, too.

Let L be the point on the line $\overline{PF_2}$ with the distance $2a$ to the focus F_2, a is the semi-major axis of the ellipse. Let line w be the bisector of the supplementary angle to the angle between the lines $\overline{PF_1}, \overline{PF_2}$. In order to prove that w is the tangent line at point P, one checks that any point Q on line w which is different from P cannot be on the ellipse. Hence w has only point P in common with the ellipse and is, therefore, the tangent at point P. From the diagram and the triangle inequality one recognizes that $2a = |LF_2| < |QF_2| + |QL| = |QF_2| + |QF_1|$ holds, which means: $|QF_2| + |QF_1| > 2a$. But if Q is a point of the ellipse, the sum should be $2a$.

Application

The rays from one focus are reflected by the ellipse to the second focus. This property has optical and acoustic applications similar to the reflective property of a parabola.

Conjugate Diameters

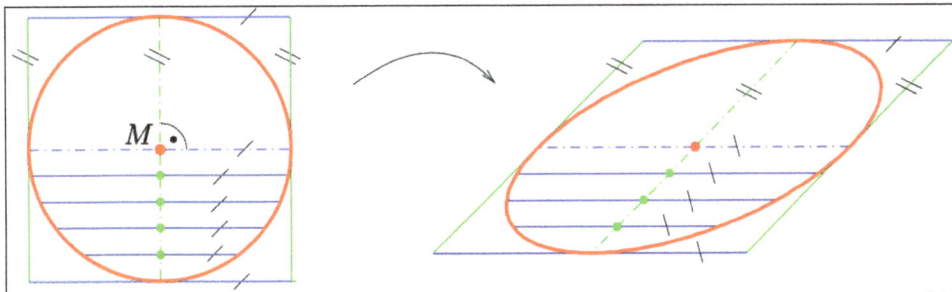

Orthogonal diameters of a circle with a square of tangents, midpoints of parallel chords and an affine image, which is an ellipse with conjugate diameters, a parallelogram of tangents and midpoints of chords

A circle has the following property:

- The midpoints of parallel chords lie on a diameter.

- An affine transformation preserves parallelism and midpoints of line segments, so this property is true for any ellipse.

- Two diameters d_1, d_2 of an ellipse are *conjugate* if the midpoints of chords parallel to d_1 lie on d_2.

From the diagram one finds:

Two diameters $\overline{P_1 Q_1}$ $\overline{P_2 Q_2}$ of an ellipse are conjugate whenever the tangents at P_1 and Q_1 are parallel to $\overline{P_2 Q_2}$.

Conjugate diameters in an ellipse generalize orthogonal diameters in a circle.

In the parametric equation for a general ellipse given above,

$$\vec{x} = \vec{p}(t) = \vec{f}_0 + \vec{f}_1 \cos t + \vec{f}_2 \sin t,$$

any pair of points $\vec{p}(t)$, $\vec{p}(t+\pi)$ belong to a diameter, and the pair $\vec{p}(t+\frac{\pi}{2})$, $\vec{p}(t-\frac{\pi}{2})$ belong to its conjugate diameter.

Theorem of Apollonios on Conjugate Diameters

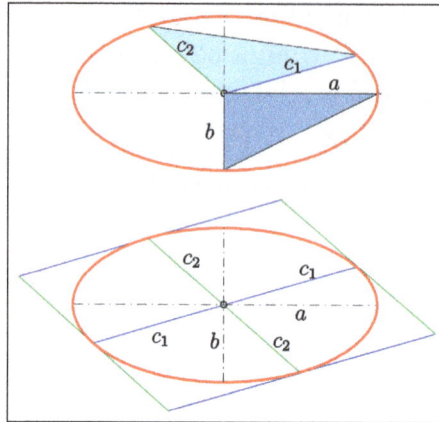

Ellipse: theorem of Apollonios on conjugate diameters.

For an ellipse with semi-axes a, b the following is true. Let c_1 and c_2 be halves of two conjugate diameters then:

1. $c_1^2 + c_2^2 = a^2 + b^2$,

2. The *triangle* formed by c_1, c_2 has the constant area $A_\Delta = \dfrac{1}{2}ab$.

3. The parallelogram of tangents adjacent to the given conjugate diameters has the $\text{Area}_{12} = 4ab$.

Let the ellipse be in the canonical form with parametric equation:

$$\vec{p}(t) = (a\cos t, b\sin t).$$

The two points $\vec{c}_1 = \vec{p}(t)$, $\vec{c}_2 = \vec{p}(t+\pi/2)$ are on conjugate diameters. From trigonometric formulae one obtains $\vec{c}_2 = (-a\sin t, b\cos t)^T$ and:

$$|\vec{c}_1|^2 + |\vec{c}_2|^2 = \cdots = a^2 + b^2 .$$

The area of the triangle generated by $\vec{c}_1, \vec{c}_2$ is:

$$A_\Delta = \tfrac{1}{2}\det(\vec{c}_1, \vec{c}_2) = \cdots = \tfrac{1}{2}ab$$

and from the diagram it can be seen that the area of the parallelogram is 8 times that of A_Δ. Hence

$$\text{Area}_{12} = 4ab .$$

Orthogonal Tangents

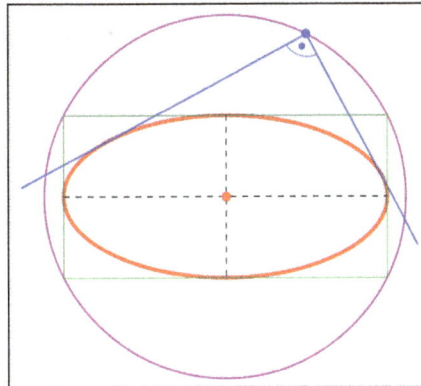

Ellipse with its orthoptic.

For the ellipse $\frac{x^2}{a^2} + \frac{y^2}{b^2} = 1$ the intersection points of *orthogonal* tangents lie on the circle $x^2 + y^2 = a^2 + b^2$.

This circle is called *orthoptic* or director circle of the ellipse.

Drawing Ellipses

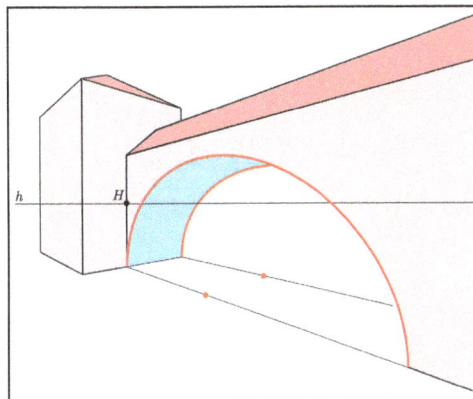

Central projection of circles (gate).

Ellipses appear in descriptive geometry as images (parallel or central projection) of circles. There exist various tools to draw an ellipse. Computers provide the fastest and most accurate method for drawing an ellipse. However, technical tools (*ellipsographs*) to draw an ellipse without a computer exist. The principle of ellipsographs were known to Greek mathematicians such as Archimedes and Proklos.

If there is no ellipsograph available, one can draw an ellipse using an approximation by the four osculating circles at the vertices.

The knowledge of the axes and the semi-axes is necessary (or equivalent: the foci and the semi-major axis).

If this presumption is not fulfilled one has to know at least two conjugate diameters. With help of Rytz's construction the axes and semi-axes can be retrieved.

De La Hire's Point Construction

The following construction of single points of an ellipse is due to de La Hire. It is based on the standard parametric representation $(a\cos t, b\sin t)$ of an ellipse:

1. Draw the two *circles* centered at the center of the ellipse with radii a, b and the axes of the ellipse.

2. Draw a *line through the center*, which intersects the two circles at point A and B, respectively.

3. The *line* through A, which is parallel to the minor axis, meets the *line* through B, which is parallel to the major axis, at an ellipse point.

4. Repeat steps (2) and (3) with different lines through the center.

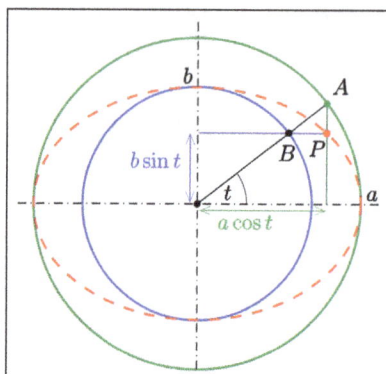

de La Hire's method.

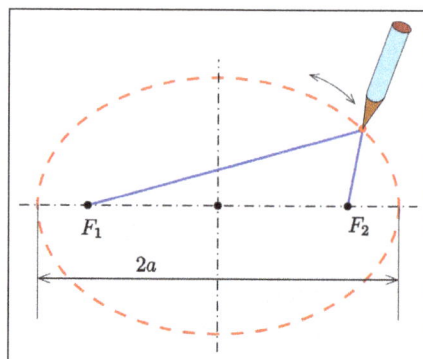

Ellipse: gardener's method.

Pins-and-string Method

The characterization of an ellipse as the locus of points so that sum of the distances to the foci is constant leads to a method of drawing one using two drawing pins, a length of string, and a pencil. In this method, pins are pushed into the paper at two points, which become the ellipse's foci. A string tied at each end to the two pins and the tip of a pencil pulls the loop taut to form a triangle. The tip of the pencil then traces an ellipse if it is moved while keeping the string taut. Using two pegs and a rope, gardeners use this procedure to outline an elliptical flower bed—thus it is called the *gardener's ellipse*.

A similar method for drawing confocal ellipses with a *closed* string is due to the Irish bishop Charles Graves.

Paper Strip Methods

The two following methods rely on the parametric representation:

$$(a\cos t, b\sin t)$$

This representation can be modeled technically by two simple methods. In both cases center, the axes and semi axes a, b have to be known.

Methods

The first method starts with:

- A strip of paper of length $a + b$.

The point, where the semi axes meet is marked by P. If the strip slides with both ends on the axes of the desired ellipse, then point P traces the ellipse. For the proof one shows that point P has the parametric representation $(a\cos t, b\sin t)$, where parameter t is the angle of the slope of the paper strip.

A technical realization of the motion of the paper strip can be achieved by a Tusi couple. The device is able to draw any ellipse with a *fixed* sum $a + b$, which is the radius of the large circle. This restriction may be a disadvantage in real life. More flexible is the second paper strip method.

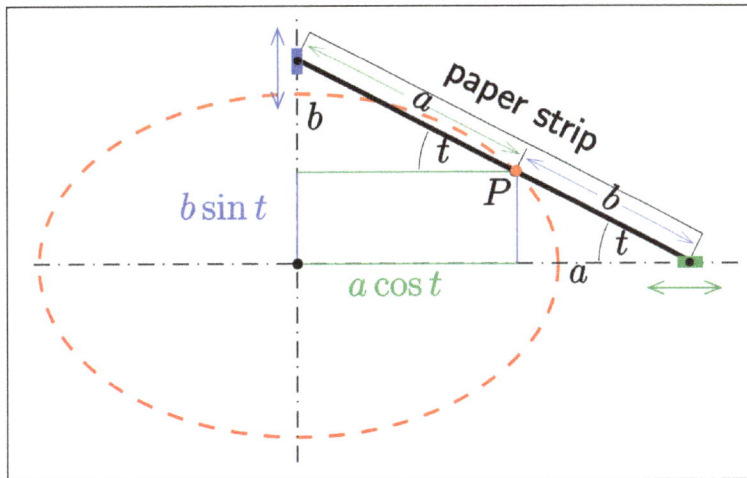

Ellipse construction: paper strip method 1.

A variation of the paper strip method 1 uses the observation that the midpoint N of the paper strip is moving on the circle with center M (of the ellipse) and radius $\frac{a+b}{2}$. Hence, the paperstrip can be cut at point N into halves, connected again by a joint at N and the sliding end K fixed at the center M. After this operation the movement of the unchanged half of the paperstrip is unchanged. This variation requires only one sliding shoe.

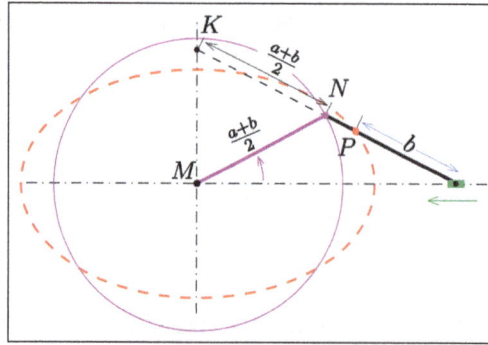

Variation of the paper strip method 1.

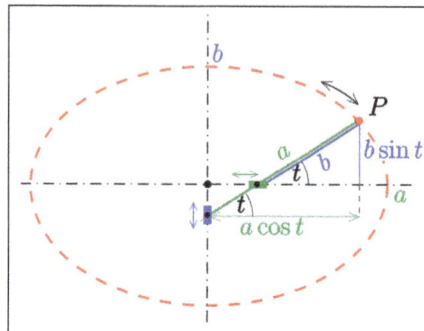

Ellipse construction: paper strip method 2.

The second method starts with:

- A strip of paper of length a.

One marks the point, which divides the strip into two substrips of length b and $a-b$. The strip is positioned onto the axes as described in the diagram. Then the free end of the strip traces an ellipse, while the strip is moved. For the proof, one recognizes that the tracing point can be described parametrically by $(a\cos t, b\sin t)$, where parameter t is the angle of slope of the paper strip.

This method is the base for several *ellipsographs*.

Similar to the variation of the paper strip method 1 a *variation of the paper strip method 2* can be established by cutting the part between the axes into halves.

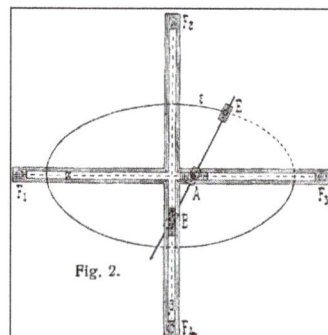

Ellipsograph due to Benjamin Bramer

Most ellipsograph drafting instruments are based on the second paperstrip method.

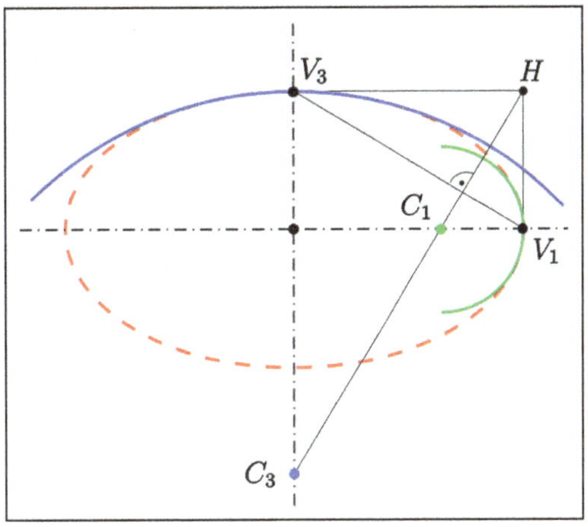

Approximation of an ellipse with osculating circles.

Approximation by Osculating Circles

From *Metric properties* below, one obtains:

- The radius of curvature at the vertices V_1, V_2 is: $\frac{b^2}{a}$

- The radius of curvature at the co-vertices V_3, V_4 is: $\frac{a^2}{b}$.

The diagram shows an easy way to find the centers of curvature $C_1 = (a - \frac{b^2}{a}, 0), C_3 = (0, b - \frac{a^2}{b})$ at vertex V_1 and co-vertex V_3, respectively:

1. mark the auxiliary point $H = (a, b)$ and draw the line segment $V_1 V_3$,

2. draw the line through H, which is perpendicular to the line $V_1 V_3$,

3. the intersection points of this line with the axes are the centers of the osculating circles.

The centers for the remaining vertices are found by symmetry.

With help of a French curve one draws a curve, which has smooth contact to the osculating circles.

Steiner Generation

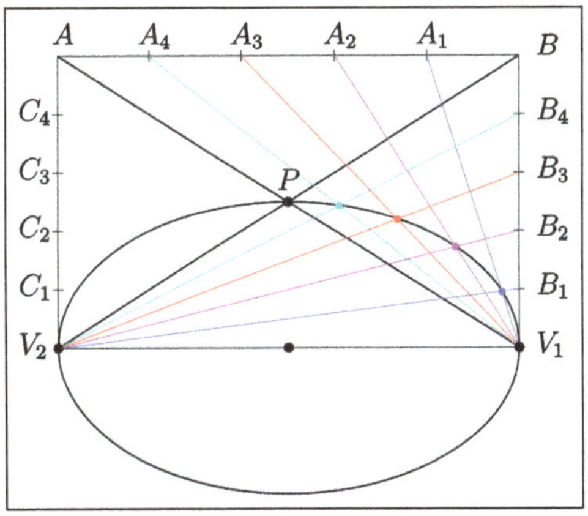

Ellipse: Steiner generation.

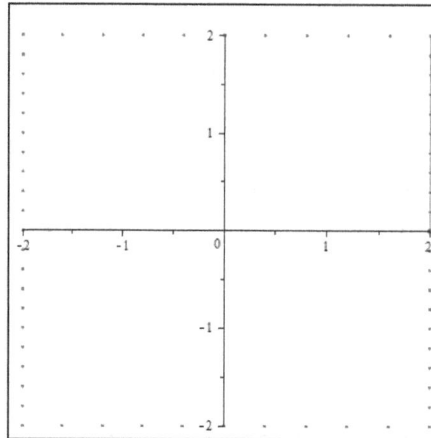

Ellipse: Steiner generation.

The following method to construct single points of an ellipse relies on the Steiner generation of a conic section:

Given two pencils $B(U), B(V)$ of lines at two points U, V (all lines containing U and V, respectively) and a projective but not perspective mapping π of $B(U)$ onto $B(V)$, then the intersection points of corresponding lines form a non-degenerate projective conic section.

For the generation of points of the ellipse $\frac{x^2}{a^2} + \frac{y^2}{b^2} = 1$ one uses the pencils at the vertices V_1, V_2. Let $P = (0, b)$ be an upper co-vertex of the ellipse and $A = (-a, 2b), B = (a, 2b)$. P is the center of the rectangle V_1, V_2, B, A. The side $\overline{AB}$ of the rectangle is divided into n equal spaced line segments and this division is projected parallel with the diagonal AV_2 as direction onto the line segment $\overline{V_1 B}$ and assign the division as shown in the diagram. The parallel projection together with the reverse of the orientation is part of the projective mapping between the pencils at V_1 and V_2 needed. The intersection points of any two related lines $V_1 B_i$ and $V_2 A_i$ are points of the uniquely defined ellipse. With help of the points $C_1, \ldots$ the points of the second quarter of the ellipse can be determined. Analogously one obtains the points of the lower half of the ellipse.

Steiner generation can also be defined for hyperbolas and parabolas. It is sometimes called a *parallelogram method* because one can use other points rather than the vertices, which starts with a parallelogram instead of a rectangle.

As Hypotrochoid

The ellipse is a special case of the hypotrochoid when $R = 2r$, as shown in the adjacent image. The special case of a moving circle with radius $R = 2r$ inside a circle with radius is called a Tusi couple.

Inscribed Angles and Three-point Form

Circles

A circle with equation $(x - x_\circ)^2 + (y - y_\circ)^2 = r^2$ is uniquely determined by three points $(x_1, y_1), (x_2, y_2), (x_3, y_3)$ not on a line. A simple way to determine the parameters $x_\circ, y_\circ, r$ uses the *inscribed angle theorem* for circles.

- For four points $P_i = (x_i, y_i)$, $i = 1, 2, 3, 4$, the following statement is true:

- The four points are on a circle if and only if the angles at P_3 and P_4 are equal.

Usually one measures inscribed angles by a degree or radian θ, but here the following measurement is more convenient:

- In order to measure the angle between two lines with equations one uses the quotient:

$$y = m_1 x + d_1, \; y = m_2 x + d_2, \; m_1 \neq m_2,$$

$$\frac{1 + m_1 \cdot m_2}{m_2 - m_1} = \cot \theta.$$

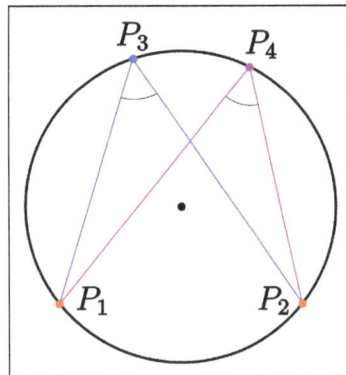

Circle: Inscribed angle theorem.

Inscribed Angle Theorem for Circles

For four points $P_i = (x_i, y_i)$, $i = 1, 2, 3, 4$, no three of them on a line, we have the following:

- The four points are on a circle, if and only if the angles at P_3 and P_4 are equal. In terms of the angle measurement above, this means:

$$\frac{(x_4 - x_1)(x_4 - x_2) + (y_4 - y_1)(y_4 - y_2)}{(y_4 - y_1)(x_4 - x_2) - (y_4 - y_2)(x_4 - x_1)} = \frac{(x_3 - x_1)(x_3 - x_2) + (y_3 - y_1)(y_3 - y_2)}{(y_3 - y_1)(x_3 - x_2) - (y_3 - y_2)(x_3 - x_1)}.$$

At first the measure is available only for chords not parallel to the y-axis, but the final formula works for any chord.

Three-point form of Circle Equation

As a consequence, one obtains an equation for the circle determined by three non-colinear points $P_i = (x_i, y_i)$:

$$\frac{(x - x_1)(x - x_2) + (y - y_1)(y - y_2)}{(y - y_1)(x - x_2) - (y - y_2)(x - x_1)} = \frac{(x_3 - x_1)(x_3 - x_2) + (y_3 - y_1)(y_3 - y_2)}{(y_3 - y_1)(x_3 - x_2) - (y_3 - y_2)(x_3 - x_1)}.$$

For example, for $P_1 = (2,0)$, $P_2 = (0,1)$, $P_3 = (0,0)$ the three-point equation is:

$$\frac{(x-2)x + y(y-1)}{yx - (y-1)(x-2)} = 0 \text{, which can be rearranged to } (x-1)^2 + (y-\tfrac{1}{2})^2 = \tfrac{5}{4}.$$

Using vectors, dot products and determinants this formula can be arranged more clearly, letting $\vec{x} = (x,y)$:

$$\frac{(\vec{x}-\vec{x}_1)\cdot(\vec{x}-\vec{x}_2)}{\det(\vec{x}-\vec{x}_1,\vec{x}-\vec{x}_2)} = \frac{(\vec{x}_3-\vec{x}_1)\cdot(\vec{x}_3-\vec{x}_2)}{\det(\vec{x}_3-\vec{x}_1,\vec{x}_3-\vec{x}_2)}.$$

The center of the circle $(x_\circ, y_\circ)$ satisfies:

$$\begin{bmatrix} 1 & \frac{y_1-y_2}{x_1-x_2} \\ \frac{x_1-x_3}{y_1-y_3} & 1 \end{bmatrix} \begin{bmatrix} x_\circ \\ y_\circ \end{bmatrix} = \begin{bmatrix} \frac{x_1^2-x_2^2+y_1^2-y_2^2}{2(x_1-x_2)} \\ \frac{y_1^2-y_3^2+x_1^2-x_3^2}{2(y_1-y_3)} \end{bmatrix}.$$

The radius is the distance between any of the three points and the center.

$$r = \sqrt{(x_1-x_\circ)^2+(y_1-y_\circ)^2} = \sqrt{(x_2-x_\circ)^2+(y_2-y_\circ)^2} = \sqrt{(x_3-x_\circ)^2+(y_3-y_\circ)^2}.$$

Ellipses

This section, we consider the family of ellipses defined by equations $\frac{(x-x_\circ)^2}{a^2} + \frac{(y-y_\circ)^2}{b^2} = 1$ with a *fixed* eccentricity e. It is convenient to use the parameter:

$$q = \frac{a^2}{b^2} = \frac{1}{1-e^2},$$

and to write the ellipse equation as:

$$(x-x_\circ)^2 + q(y-y_\circ)^2 = a^2,$$

where q is fixed and $x_\circ, y_\circ, a$ vary over the real numbers. (Such ellipses have their axes parallel to the coordinate axes: if $q<1$, the major axis is parallel to the x-axis; if $q>1$, it is parallel to the y-axis.)

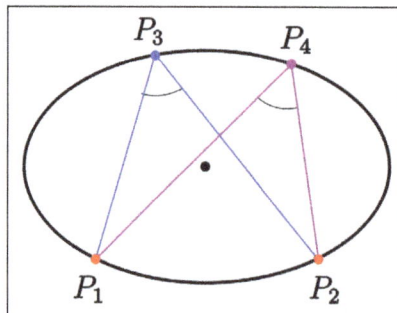

Inscribed angle theorem for an ellipse.

Like a circle, such an ellipse is determined by three points not on a line.

For this family of ellipses, one introduces the following q-analog angle measure, which is *not* a function of the usual angle measure θ:

- In order to measure an angle between two lines with equations one uses the quotient:

$$y = m_1 x + d_1, \, y = m_2 x + d_2 \,, m_1 \neq m_2$$

$$\frac{1 + q \, m_1 \cdot m_2}{m_2 - m_1}.$$

Inscribed Angle Theorem for Ellipses

Given four points $P_i = (x_i, y_i), i = 1, 2, 3, 4, ,$ no three of them on a line.

The four points are on an ellipse with equation $(x - x_\circ)^2 + q(y - y_\circ)^2 = a^2$ if and only if the angles at P_3 and P_4 are equal in the sense of the measurement above—that is, if:

$$\frac{(x_4 - x_1)(x_4 - x_2) + q\,(y_4 - y_1)(y_4 - y_2)}{(y_4 - y_1)(x_4 - x_2) - (y_4 - y_2)(x_4 - x_1)} = \frac{(x_3 - x_1)(x_3 - x_2) + q\,(y_3 - y_1)(y_3 - y_2)}{(y_3 - y_1)(x_3 - x_2) - (y_3 - y_2)(x_3 - x_1)}.$$

At first the measure is available only for chords which are not parallel to the y-axis. But the final formula works for any chord. The proof follows from a straightforward calculation. For the direction of proof given that the points are on an ellipse, one can assume that the center of the ellipse is the origin.

Three-point Form of Ellipse Equation

A consequence, one obtains an equation for the ellipse determined by three non-colinear points $P_i = (x_i, y_i)$:

$$\frac{(x - x_1)(x - x_2) + q\,(y - y_1)(y - y_2)}{(y - y_1)(x - x_2) - (y - y_2)(x - x_1)} = \frac{(x_3 - x_1)(x_3 - x_2) + q\,(y_3 - y_1)(y_3 - y_2)}{(y_3 - y_1)(x_3 - x_2) - (y_3 - y_2)(x_3 - x_1)}.$$

For example, for $P_1 = (2, 0)$, $P_2 = (0, 1)$, $P_3 = (0, 0)$ and $q = 4$ one obtains the three-point form:

$$\frac{(x - 2)x + 4y(y - 1)}{yx - (y - 1)(x - 2)} = 0 \quad \text{and after conversion} \quad \frac{(x - 1)^2}{2} + \frac{(y - \frac{1}{2})^2}{1/2} = 1.$$

Analogously to the circle case, the equation can be written more clearly using vectors:

$$\frac{(\vec{x} - \vec{x}_1) * (\vec{x} - \vec{x}_2)}{\det(\vec{x} - \vec{x}_1, \vec{x} - \vec{x}_2)} = \frac{(\vec{x}_3 - \vec{x}_1) * (\vec{x}_3 - \vec{x}_2)}{\det(\vec{x}_3 - \vec{x}_1, \vec{x}_3 - \vec{x}_2)},$$

where $*$ is the modified dot product $\vec{u} * \vec{v} = u_x v_x + q u_y v_y$.

Pole-polar Relation

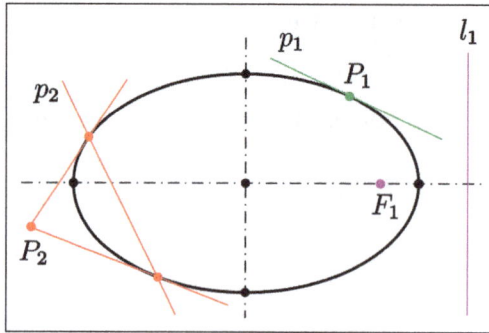

Ellipse: pole-polar relation.

Any ellipse can be described in a suitable coordinate system by an equation $\frac{x^2}{a^2}+\frac{y^2}{b^2}=1$. The equation of the tangent at a point $P_1=(x_1,y_1)$ of the ellipse is $\frac{x_1x}{a^2}+\frac{y_1y}{b^2}=1$. If one allows point $P_1=(x_1,y_1)$ to be an arbitrary point different from the origin, then:

- Point $P_1=(x_1,y_1)\neq(0,0)$ is mapped onto the line $\frac{x_1x}{a^2}+\frac{y_1y}{b^2}=1$, not through the center of the ellipse.

This relation between points and lines is a bijection.

The inverse function maps:

- Line $y=mx+d, d\neq0$ onto the point $\left(-\frac{ma^2}{d},\frac{b^2}{d}\right)$.
- Line $x=c, c\neq0$ onto the point $\left(\frac{a^2}{c},0\right)$.

Such a relation between points and lines generated by a conic is called *pole-polar relation* or *polarity*. The pole is the point, the polar the line.

By calculation one can confirm the following properties of the pole-polar relation of the ellipse:

- For a point (pole) *on* the ellipse the polar is the tangent at this point.

- For a pole *outside* the ellipse the intersection points of its polar with the ellipse are the tangency points of the two tangents passing P .sss

- For a point *within* the ellipse the polar has no point with the ellipse in common..

1. The intersection point of two polars is the pole of the line through their poles.

2. The foci $(c,0)$, and $(-c,0)$ respectively and the directrices $x=\frac{a^2}{c}$ and $x=-\frac{a^2}{c}$ respectively belong to pairs of pole and polar.

Pole-polar relations exist for hyperbolas and parabolas, too.

Metric Properties

All metric properties given below refer to an ellipse with equation $\dfrac{x^2}{a^2}+\dfrac{y^2}{b^2}=1$.

Area

The area A_{ellipse} enclosed by an ellipse is:

- $A_{\text{ellipse}} = \pi ab$

where a and b are the lengths of the semi-major and semi-minor axes, respectively. The area formula πab is intuitive: start with a circle of radius b (so its area is πb^2) and stretch it by a factor to make an ellipse. This scales the area by the same factor: $\pi b^2 (a/b) = \pi ab$. It is also easy to rigorously prove the area formula using integration as follows. Equation (1) can be rewritten as $y(x) = b\sqrt{1 - x^2/a^2}$. For $x \in [-a, a]$, this curve is the top half of the ellipse. So twice the integral of $y(x)$ over the interval $[-a, a]$ will be the area of the ellipse:

$$A_{\text{ellipse}} = \int_{-a}^{a} 2b\sqrt{1 - x^2/a^2}\, dx$$

$$= \frac{b}{a} \int_{-a}^{a} 2\sqrt{a^2 - x^2}\, dx.$$

The second integral is the area of a circle of radius a that is, πa^2. So,

$$A_{\text{ellipse}} = \frac{b}{a} \pi a^2 = \pi ab.$$

An ellipse defined implicitly by $Ax^2 + Bxy + Cy^2 = 1$ has area $2\pi / \sqrt{4AC - B^2}$.

The area can also be expressed in terms of eccentricity and the length of the semi-major axis as $a^2 \pi \sqrt{1 - e^2}$ (obtained by solving for flattening, then computing the semi-minor axis).

Circumference

The circumference C of an ellipse is:

$$C = 4a \int_{0}^{\pi/2} \sqrt{1 - e^2 \sin^2 \theta}\, d\theta = 4aE(e)$$

where again a is the length of the semi-major axis, $e = \sqrt{1 - b^2/a^2}$ is the eccentricity, and the function E is the complete elliptic integral of the second kind,

$$E(e) = \int_{0}^{\pi/2} \sqrt{1 - e^2 \sin^2 \theta}\, d\theta.$$

The circumference of the ellipse may be evaluated in terms of $E(e)$ using Gauss's arithmetic-geometric mean; this is a quadratically converging iterative method.

The exact infinite series is:

$$C = 2\pi a\left[1-\left(\frac{1}{2}\right)^2 e^2 - \left(\frac{1\cdot 3}{2\cdot 4}\right)^2 \frac{e^4}{3} - \left(\frac{1\cdot 3\cdot 5}{2\cdot 4\cdot 6}\right)^2 \frac{e^6}{5} - \cdots\right]$$

$$= 2\pi a\left[1-\sum_{n=1}^{\infty}\left(\frac{(2n-1)!!}{(2n)!!}\right)^2 \frac{e^{2n}}{2n-1}\right],$$

where $n!!$ is the double factorial. This series converges, but by expanding in terms of $h = (a-b)^2 / (a+b)^2$, James Ivory and Bessel derived an expression that converges much more rapidly:

$$C = \pi(a+b)\left[1+\sum_{n=1}^{\infty}\left(\frac{(2n-1)!!}{2^n n!}\right)^2 \frac{h^n}{(2n-1)^2}\right].$$

$$= \pi(a+b)\left[1+\frac{h}{4}+\sum_{n=2}^{\infty}\left(\frac{(2n-3)!!}{2^n n!}\right)^2 h^n\right].$$

Srinivasa Ramanujan gives two close approximations for the circumference in §16 of "Modular Equations and Approximations to π"; they are:

$$C \approx \pi\left[3(a+b)-\sqrt{(3a+b)(a+3b)}\right] = \pi\left[3(a+b)-\sqrt{10ab+3(a^2+b^2)}\right]$$

and,

$$C \approx \pi(a+b)\left(1+\frac{3h}{10+\sqrt{4-3h}}\right).$$

The errors in these approximations, which were obtained empirically, are of order h^3 and h^5, respectively.

More generally, the arc length of a portion of the circumference, as a function of the angle subtended (or x-coordinates of any two points on the upper half of the ellipse), is given by an incomplete elliptic integral. The upper half of an ellipse is parameterized by:

$$y = b\sqrt{1-\frac{x^2}{a^2}}.$$

Then the arc length s from x_1 to x_2 is:

$$s = -b\int_{\arccos\frac{x_1}{a}}^{\arccos\frac{x_2}{a}}\sqrt{1-\left(1-\frac{a^2}{b^2}\right)\sin^2 z}\, dz.$$

This is equivalent to,

$$s = -b\left[E\left(z\Big|1-\frac{a^2}{b^2}\right)\right]_{\arccos\frac{x_1}{a}}^{\arccos\frac{x_2}{a}}$$

where $E(z|m)$ is the incomplete elliptic integral of the second kind with parameter $m = k^2$.

The inverse function, the angle subtended as a function of the arc length, is given by a certain elliptic function.

Some lower and upper bounds on the circumference of the canonical ellipse $x^2 / a^2 + y^2 / b^2 = 1$ with $a \geq b$ are:

$$2\pi b \leq C \leq 2\pi a,$$

$$\pi(a+b) \leq C \leq 4(a+b),$$

$$4\sqrt{a^2 + b^2} \leq C \leq \sqrt{2}\pi\sqrt{a^2 + b^2}.$$

Here the upper bound $2\pi a$ is the circumference of a circumscribed concentric circle passing through the endpoints of the ellipse's major axis, and the lower bound $4\sqrt{a^2 + b^2}$ is the perimeter of an inscribed rhombus with vertices at the endpoints of the major and the minor axes.

Curvature

The curvature is given by $\kappa = \frac{1}{a^2 b^2}\left(\frac{x^2}{a^4} + \frac{y^2}{b^4}\right)^{-\frac{3}{2}}$, radius of curvature at point (x, y):

$$\rho = a^2 b^2 \left(\frac{x^2}{a^4} + \frac{y^2}{b^4}\right)^{3/2} = \frac{1}{a^4 b^4}\sqrt{\left(a^4 y^2 + b^4 x^2\right)^3}.$$

Radius of curvature at the two *vertices* $(\pm a, 0)$ and the centers of curvature:

$$\rho_0 = \frac{b^2}{a} = p, \qquad \left(\pm\frac{c^2}{a}\Big|0\right).$$

Radius of curvature at the two *co-vertices* $(0, \pm b)$ and the centers of curvature:

$$\rho_1 = \frac{a^2}{b}, \qquad \left(0\Big|\pm\frac{c^2}{b}\right).$$

In Triangle Geometry

Ellipses appear in triangle geometry as:

1.   Steiner ellipse: Ellipse through the vertices of the triangle with center at the centroid.

2. Inellipses: ellipses which touch the sides of a triangle. Special cases are the Steiner inellipse and the Mandart inellipse.

Applications

Elliptical Reflectors and Acoustics

If the water's surface is disturbed at one focus of an elliptical water tank, the circular waves of that disturbance, after reflecting off the walls, converge simultaneously to a single point: the *second focus*. This is a consequence of the total travel length being the same along any wall-bouncing path between the two foci.

Similarly, if a light source is placed at one focus of an elliptic mirror, all light rays on the plane of the ellipse are reflected to the second focus. Since no other smooth curve has such a property, it can be used as an alternative definition of an ellipse. (In the special case of a circle with a source at its center all light would be reflected back to the center.) If the ellipse is rotated along its major axis to produce an ellipsoidal mirror (specifically, a prolate spheroid), this property holds for all rays out of the source. Alternatively, a cylindrical mirror with elliptical cross-section can be used to focus light from a linear fluorescent lamp along a line of the paper; such mirrors are used in some document scanners.

Sound waves are reflected in a similar way, so in a large elliptical room a person standing at one focus can hear a person standing at the other focus remarkably well. The effect is even more evident under a vaulted roof shaped as a section of a prolate spheroid. Such a room is called a *whisper chamber*. The same effect can be demonstrated with two reflectors shaped like the end caps of such a spheroid, placed facing each other at the proper distance. Examples are the National Statuary Hall at the United States Capitol (where John Quincy Adams is said to have used this property for eavesdropping on political matters); the Mormon Tabernacle at Temple Square in Salt Lake City, Utah; at an exhibit on sound at the Museum of Science and Industry in Chicago; in front of the University of Illinois at Urbana–Champaign Foellinger Auditorium; and also at a side chamber of the Palace of Charles V, in the Alhambra.

Planetary Orbits

In the 17th century, Johannes Kepler discovered that the orbits along which the planets travel around the Sun are ellipses with the Sun [approximately] at one focus, in his first law of planetary motion. Later, Isaac Newton explained this as a corollary of his law of universal gravitation.

More generally, in the gravitational two-body problem, if the two bodies are bound to each other (that is, the total energy is negative), their orbits are similar ellipses with the common barycenter being one of the foci of each ellipse. The other focus of either ellipse has no known physical significance. The orbit of either body in the reference frame of the other is also an ellipse, with the other body at the same focus.

Keplerian elliptical orbits are the result of any radially directed attraction force whose strength is inversely proportional to the square of the distance. Thus, in principle, the motion of two oppositely charged particles in empty space would also be an ellipse. (However, this conclusion ignores losses due to electromagnetic radiation and quantum effects, which become significant when the particles are moving at high speed.)

For elliptical orbits, useful relations involving the eccentricity e are:

$$e = \frac{r_a - r_p}{r_a + r_p} = \frac{r_a - r_p}{2a}$$

$$r_a = (1+e)a$$

$$r_p = (1-e)a$$

where,

- r_a is the radius at apoapsis (the farthest distance).

- r_p is the radius at periapsis (the closest distance).

- a is the length of the semi-major axis.

Also, in terms of r_a and r_p, the semi-major axis a is their arithmetic mean, the semi-minor axis b is their geometric mean, and the semi-latus rectum l is their harmonic mean. In other words,

$$a = \frac{r_a + r_p}{2}$$

$$b = \sqrt{r_a \cdot r_p} \ .$$

$$\ell = \frac{2}{\dfrac{1}{r_a} + \dfrac{1}{r_p}} = \frac{2 r_a r_p}{r_a + r_p} \ .$$

Harmonic Oscillators

The general solution for a harmonic oscillator in two or more dimensions is also an ellipse. Such is the case, for instance, of a long pendulum that is free to move in two dimensions; of a mass attached to a fixed point by a perfectly elastic spring; or of any object that moves under influence of an attractive force that is directly proportional to its distance from a fixed attractor. Unlike Keplerian orbits, however, these "harmonic orbits" have the center of attraction at the geometric center of the ellipse, and have fairly simple equations of motion.

Phase Visualization

In electronics, the relative phase of two sinusoidal signals can be compared by feeding them to the vertical and horizontal inputs of an oscilloscope. If the Lissajous figure display is an ellipse, rather than a straight line, the two signals are out of phase.

Elliptical Gears

Two non-circular gears with the same elliptical outline, each pivoting around one focus and positioned at the proper angle, turn smoothly while maintaining contact at all times. Alternatively, they can be connected by a link chain or timing belt, or in the case of a bicycle the main chainring may be elliptical, or an ovoid similar to an ellipse in form. Such elliptical gears may be used in mechanical equipment to produce variable angular speed or torque from a constant rotation of the driving axle, or in the case of a bicycle to allow a varying crank rotation speed with inversely varying mechanical advantage.

Elliptical bicycle gears make it easier for the chain to slide off the cog when changing gears.

An example gear application would be a device that winds thread onto a conical bobbin on a spinning machine. The bobbin would need to wind faster when the thread is near the apex than when it is near the base.

Optics

- In a material that is optically anisotropic (birefringent), the refractive index depends on the direction of the light. The dependency can be described by an index ellipsoid. (If the material is optically isotropic, this ellipsoid is a sphere).

- In lamp-pumped solid-state lasers, elliptical cylinder-shaped reflectors have been used to direct light from the pump lamp (coaxial with one ellipse focal axis) to the active medium rod (coaxial with the second focal axis).

- In laser-plasma produced EUV light sources used in microchip lithography, EUV light is generated by plasma positioned in the primary focus of an ellipsoid mirror and is collected in the secondary focus at the input of the lithography machine.

Statistics and Finance

In statistics, a bivariate random vector (X, Y) is jointly elliptically distributed if its iso-density contours—loci of equal values of the density function—are ellipses. The concept extends to an arbitrary number of elements of the random vector, in which case in general the iso-density contours are ellipsoids. A special case is the multivariate normal distribution. The elliptical distributions are important in finance because if rates of return on assets are jointly elliptically distributed then all portfolios can be characterized completely by their mean and variance—that is, any two portfolios with identical mean and variance of portfolio return have identical distributions of portfolio return.

Computer Graphics

Drawing an ellipse as a graphics primitive is common in standard display libraries, such as the MacIntosh QuickDraw API, and Direct2D on Windows. Jack Bresenham at IBM is most famous for the invention of 2D drawing primitives, including line and circle drawing, using only fast integer operations such as addition and branch on carry bit. M. L. V. Pitteway extended Bresenham's algorithm for lines to conics in 1967. Another efficient generalization to draw ellipses was invented in 1984 by Jerry Van Aken.

In 1970 Danny Cohen presented at the "Computer Graphics 1970" conference in England a linear algorithm for drawing ellipses and circles. In 1971, L. B. Smith published similar algorithms for all conic sections and proved them to have good properties. These algorithms need only a few multiplications and additions to calculate each vector.

It is beneficial to use a parametric formulation in computer graphics because the density of points is greatest where there is the most curvature. Thus, the change in slope between each successive point is small, reducing the apparent "jaggedness" of the approximation.

Drawing with Bézier Paths

Composite Bézier curves may also be used to draw an ellipse to sufficient accuracy, since any ellipse may be construed as an affine transformation of a circle. The spline methods used to draw a circle may be used to draw an ellipse, since the constituent Bézier curves behave appropriately under such transformations.

Optimization Theory

It is sometimes useful to find the minimum bounding ellipse on a set of points. The ellipsoid method is quite useful for attacking this problem.

References

- Sondow, Jonathan (2013). "The parbelos, a parabolic analog of the arbelos". American Mathematical Monthly. 120: 929–935. Arxiv:1210.2279. Doi:10.4169/amer.math.monthly.120.10.929.

- Conic-section, science: britannica.com, Retrieved 10 July, 2019

- Troyano, Leonardo Fernández (2003). Bridge engineering: a global perspective. Thomas Telford. P. 536. ISBN 0-7277-3215-3.

- Circles, mensuration, +maths, guides: toppr.com, Retrieved 22 May, 2019

- Fitzpatrick, Richard (July 14, 2007). "Spherical Mirrors". Electromagnetism and Optics, lectures. University of Texas at Austin. Paraxial Optics. Retrieved October 5, 2011.

- Circle-mathematics, science: britannica.com, Retrieved 23 February, 2019

6

Euclidean Geometry

Euclidean geometry refers to the study and analysis of solid shapes, figures and planes, on the basis of axioms, postulates and theorems given by Greek mathematician Euclid. This chapter delves into detailed study of the axioms of Euclidean plane geometry, Euclid's postulates, etc. to provide in-depth understanding of Euclidean geometry.

Euclidean geometry is the study of plane and solid figures on the basis of axioms and theorems employed by the Greek mathematician Euclid. In its rough outline, Euclidean geometry is the plane and solid geometry commonly taught in secondary schools. Indeed, until the second half of the 19th century, when non-Euclidean geometries attracted the attention of mathematicians, geometry meant Euclidean geometry. It is the most typical expression of general mathematical thinking. Rather than the memorization of simple algorithms to solve equations by rote, it demands true insight into the subject, clever ideas for applying theorems in special situations, an ability to generalize from known facts, and an insistence on the importance of proof. In Euclid's great work, the Elements, the only tools employed for geometrical constructions were the ruler and the compass—a restriction retained in elementary Euclidean geometry to this day.

In its rigorous deductive organization, the Elements remained the very model of scientific exposition until the end of the 19th century, when the German mathematician David Hilbert wrote his famous Foundations of Geometry. The modern version of Euclidean geometry is the theory of Euclidean (coordinate) spaces of multiple dimensions, where distance is measured by a suitable generalization of the Pythagorean theorem.

Fundamentals of Euclidean Geometry

Euclid realized that a rigorous development of geometry must start with the foundations. Hence, he began the Elements with some undefined terms, such as "a point is that which has no part" and "a line is a length without breadth." Proceeding from these terms, he defined further ideas such as angles, circles, triangles, and various other polygons and figures. For example, an angle was defined as the inclination of two straight lines, and a circle was a plane figure consisting of all points that have a fixed distance (radius) from a given centre.

As a basis for further logical deductions, Euclid proposed five common notions, such as "things equal to the same thing are equal," and five unprovable but intuitive principles known variously as postulates or axioms. Stated in modern terms, the axioms are as follows:

- Given two points, there is a straight line that joins them.

- A straight line segment can be prolonged indefinitely.

- A circle can be constructed when a point for its centre and a distance for its radius are given.

- All right angles are equal.

- If a straight line falling on two straight lines makes the interior angles on the same side less than two right angles, the two straight lines, if produced indefinitely, will meet on that side on which the angles are less than the two right angles.

The fifth axiom became known as the "parallel postulate," since it provided a basis for the uniqueness of parallel lines. (It also attracted great interest because it seemed less intuitive or self-evident than the others. In the 19th century, Carl Friedrich Gauss, János Bolyai, and Nikolay Lobachevsky all began to experiment with this postulate, eventually arriving at new, non-Euclidean, geometries.) All five axioms provided the basis for numerous provable statements, or theorems, on which Euclid built his geometry.

Plane Geometry

Congruence of Triangles

Two triangles are said to be congruent if one can be exactly superimposed on the other by a rigid motion, and the congruence theorems specify the conditions under which this can occur. The first such theorem is the side-angle-side (SAS) theorem: If two sides and the included angle of one triangle are equal to two sides and the included angle of another triangle, the triangles are congruent. Following this, there are corresponding angle-side-angle (ASA) and side-side-side (SSS) theorems.

The first very useful theorem derived from the axioms is the basic symmetry property of isosceles triangles—i.e., that two sides of a triangle are equal if and only if the angles opposite them are equal. Euclid's proof of this theorem was once called Pons Asinorum ("Bridge of Asses"), supposedly because mediocre students could not proceed across it to the farther reaches of geometry. The Bridge of Asses opens the way to various theorems on the congruence of triangles.

The parallel postulate is fundamental for the proof of the theorem that the sum of the angles of a triangle is always 180 degrees. A simple proof of this theorem was attributed to the Pythagoreans.

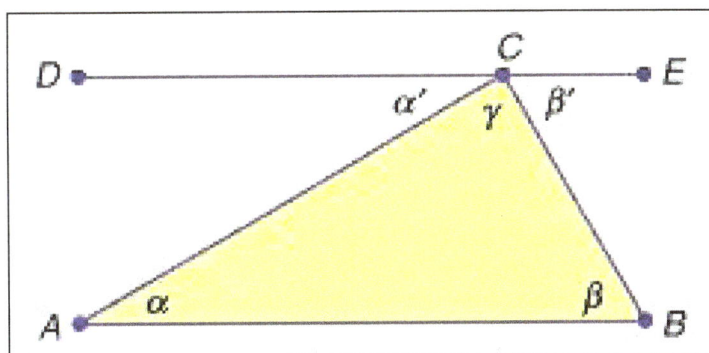

Proof that the sum of the angles in a triangle is 180 degrees. According to an ancient theorem, a transversal through two parallel lines (DE and AB in the figure) forms several equal angles, such as the alternating angles α/α' and β/β', labeled in the figure. By definition, the three angles α', γ, and β' on the line DE must sum to 180 degrees. Since $\alpha = \alpha'$ and $\beta = \beta'$, the sum of the angles in the triangle (α, β, and γ) is also 180 degrees.

Similarity of Triangles

As indicated above, congruent figures have the same shape and size. Similar figures, on the other hand, have the same shape but may differ in size. Shape is intimately related to the notion of proportion, as ancient Egyptian artisans observed long ago. Segments of lengths a, b, c, and d are said to be proportional if a:b = c:d (read, a is to b as c is to d; in older notation a:b::c:d). The fundamental theorem of similarity states that a line segment splits two sides of a triangle into proportional segments if and only if the segment is parallel to the triangle's third side.

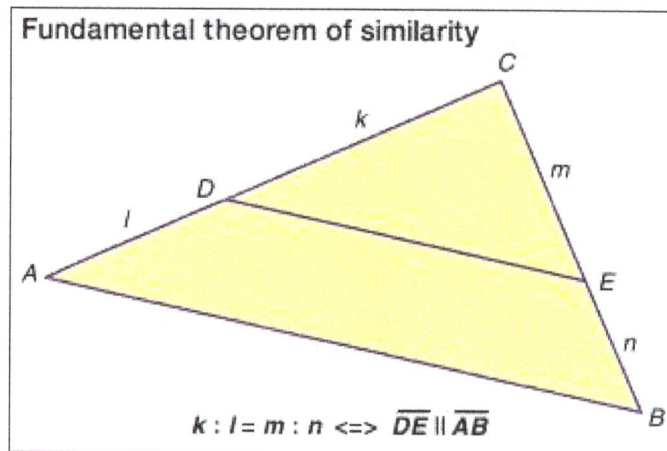

The formula in the figure reads k is to l as m is to n if and only if line DE is parallel to line AB.
This theorem then enables one to show that the small and large triangles are similar.

The similarity theorem may be reformulated as the AAA (angle-angle-angle) similarity theorem: two triangles have their corresponding angles equal if and only if their corresponding sides are proportional. Two similar triangles are related by a scaling (or similarity) factor s: if the first triangle has sides a, b, and c, then the second one will have sides sa, sb, and sc. In addition to the ubiquitous use of scaling factors on construction plans and geographic maps, similarity is fundamental to trigonometry.

Areas

Just as a segment can be measured by comparing it with a unit segment, the area of a polygon or other plane figure can be measured by comparing it with a unit square. The common formulas for calculating areas reduce this kind of measurement to the measurement of certain suitable lengths. The simplest case is a rectangle with sides a and b, which has area ab. By putting a triangle into an appropriate rectangle, one can show that the area of the triangle is half the product of the length of one of its bases and its corresponding height—bh/2. One can then compute the area of a general polygon by dissecting it into triangular regions. If a triangle has area A, a similar triangle sswith a scaling factor of s will have an area of s^2A.

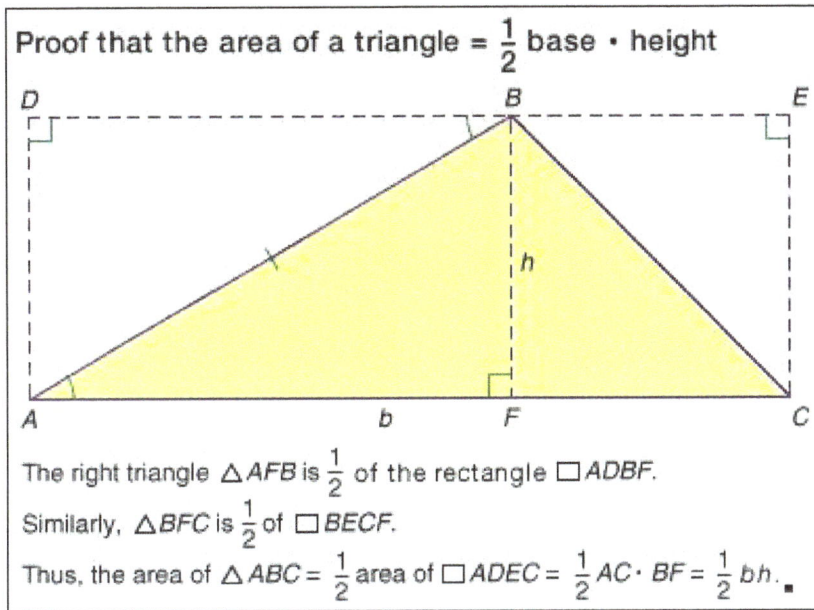

The right triangle △AFB is $\frac{1}{2}$ of the rectangle ▭ADBF.

Similarly, △BFC is $\frac{1}{2}$ of ▭BECF.

Thus, the area of △ABC = $\frac{1}{2}$ area of ▭ADEC = $\frac{1}{2}$ AC · BF = $\frac{1}{2}$ bh. ∎

Area of a triangle.

Pythagorean Theorem

For a triangle △ABC the Pythagorean theorem has two parts: (1) if ∠ACB is a right angle, then $a^2 + b^2 = c^2$; (2) if $a^2 + b^2 = c^2$, then ∠ACB is a right angle. For an arbitrary triangle, the Pythagorean theorem is generalized to the law of cosines: $a^2 + b^2 = c^2 - 2ab \cos (\angle ACB)$. When ∠ACB is 90 degrees, this reduces to the Pythagorean theorem because $\cos (90°) = 0$.

Since Euclid, a host of professional and amateur mathematicians have found more than 300 distinct proofs of the Pythagorean theorem. Despite its antiquity, it remains one of the most important theorems in mathematics. It enables one to calculate distances or, more important, to define distances in situations far more general than elementary geometry. For example, it has been generalized to multidimensional vector spaces.

Circles

A chord AB is a segment in the interior of a circle connecting two points (A and B) on the circumference. When a chord passes through the circle's centre, it is a diameter, d. The circumference of a circle is given by πd, or 2πr where r is the radius of the circle; the area of a circle is πr^2. In each case, π is the same constant. Mathematician Archimedes used the method of exhaustion to obtain upper and lower bounds for π by circumscribing and inscribing regular polygons about a circle.

A semicircle has its end points on a diameter of a circle. Thales (flourished 6th century BCE) is generally credited with having proved that any angle inscribed in a semicircle is a right angle; that is, for any point C on the semicircle with diameter AB, ∠ACB will always be 90 degrees. Another important theorem states that for any chord AB in a circle, the angle subtended by any point on the same semiarc of the circle will be invariant. Slightly modified, this means that in a circle, equal chords determine equal angles, and vice versa.

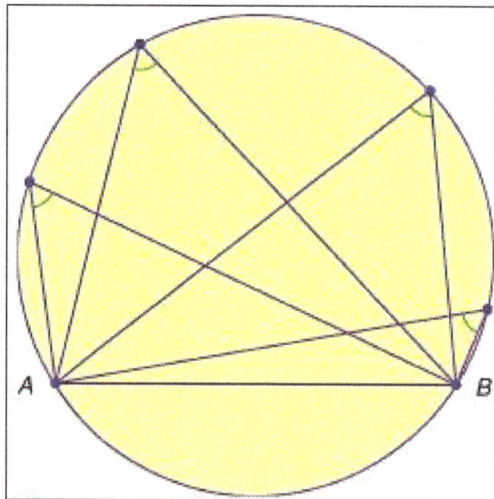

Thales of Miletus is generally credited with giving the first proof that for any chord AB in a circle, all of the angles subtended by points anywhere on the same semiarc of the circle will be equal.

Summarizing the above material, the five most important theorems of plane Euclidean geometry are: the sum of the angles in a triangle is 180 degrees, the Bridge of Asses, the fundamental theorem of similarity, the Pythagorean theorem, and the invariance of angles subtended by a chord in a circle. Most of the more advanced theorems of plane Euclidean geometry are proved with the help of these theorems.

Regular Polygons

A polygon is called regular if it has equal sides and angles. Thus, a regular triangle is an equilateral triangle, and a regular quadrilateral is a square. A general problem since antiquity has been the problem of constructing a regular n-gon, for different n, with only ruler and compass. For example, Euclid constructed a regular pentagon by applying the above-mentioned five important theorems in an ingenious combination.

Techniques, such as bisecting the angles of known constructions, exist for constructing regular n-gons for many values, but none is known for the general case. In 1797, following centuries without any progress, Gauss surprised the mathematical community by discovering a construction for the 17-gon. More generally, Gauss was able to show that for a prime number p, the regular p-gon is constructible if and only if p is a "Fermat prime": $p = F(k) = 2^{2^k} + 1$. Because it is not known in general which $F(k)$ are prime, the construction problem for regular n-gons is still open.

Three other unsolved construction problems from antiquity were finally settled in the 19th century by applying tools not available to the Greeks. Comparatively simple algebraic methods showed that it is not possible to trisect an angle with ruler and compass or to construct a cube with a volume double that of a given cube. Showing that it is not possible to square a circle (i.e., to construct a square equal in area to a given circle by the same means), however, demanded deeper insights into the nature of the number π.

Solid Geometry

The most important difference between plane and solid Euclidean geometry is that human beings can look at the plane "from above," whereas three-dimensional space cannot be looked at "from outside". Consequently, intuitive insights are more difficult to obtain for solid geometry than for plane geometry.

Some concepts, such as proportions and angles, remain unchanged from plane to solid geometry. For other familiar concepts, there exist analogies—most noticeably, volume for area and three-dimensional shapes for two-dimensional shapes (sphere for circle, tetrahedron for triangle, box for rectangle). However, the theory of tetrahedra is not nearly as rich as it is for triangles. Active research in higher-dimensional Euclidean geometry includes convexity and sphere packings and their applications in cryptology and crystallography.

Volume

As explained above, in plane geometry the area of any polygon can be calculated by dissecting it into triangles. A similar procedure is not possible for solids. In 1901 the German mathematician Max Dehn showed that there exist a cube and a tetrahedron of equal volume that cannot be dissected and rearranged into each other. This means that calculus must be used to calculate volumes for even many simple solids such as pyramids.

Regular Solids

Regular polyhedra are the solid analogies to regular polygons in the plane. Regular polygons are defined as having equal (congruent) sides and angles. In analogy, a solid is called regular if its faces are congruent regular polygons and its polyhedral angles (angles at which the faces meet) are congruent. This concept has been generalized to higher-dimensional (coordinate) Euclidean spaces.

Whereas in the plane there exist (in theory) infinitely many regular polygons, in three-dimensional space there exist exactly five regular polyhedra. These are known as the Platonic solids: the tetrahedron, or pyramid, with 4 triangular faces; the cube, with 6 square faces; the octahedron, with 8 equilateral triangular faces; the dodecahedron, with 12 pentagonal faces; and the icosahedron, with 20 equilateral triangular faces.

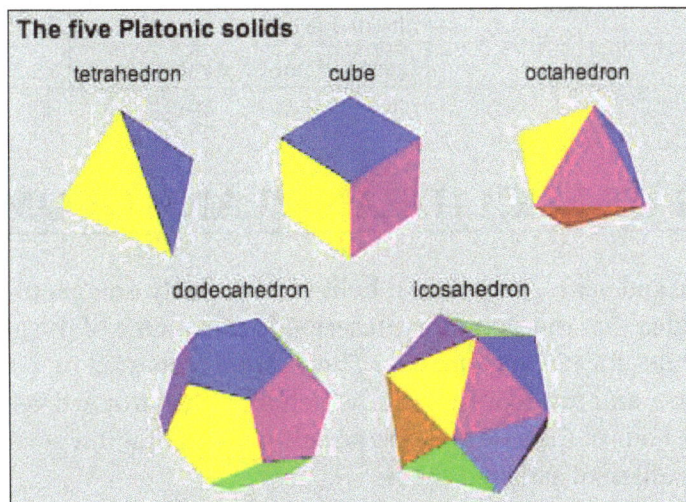

The five Platonic solidsThese are the only geometric solids whose faces are composed
of regular, identical polygons. Placing the cursor on each figure will show it in animation.

In four-dimensional space there exist exactly six regular polytopes, five of them generalizations from three-dimensional space. In any space of more than four dimensions, there exist exactly three regular polytopes—the generalizations of the tetrahedron, the cube, and the octahedron.

Calculating Areas and Volumes

The table presents mathematical formulas for calculating the areas of various plane figures and the volumes of various solid figures.

Mathematical Formulas

	Shape	Action	Formula
Circumference	Circle	Multiply diameter by π	πd
Area	Circle	Multiply radius squared by π	πr^2
	Rectangle	Multiply height by length	hl
	Sphere surface	Multiply radius squared by π by 4	$4\pi r^2$
	Square	Length of one side squared	s^2
	Trapezoid	Parallel side length A + parallel side length B multiplied by height and divided by 2	$(A + B)h/2$
	Triangle	Multiply base by height and divide by 2	$hb/2$
Volume	cone	Multiply base radius squared by π by height and divide by 3	$br^2\pi h/3$
	Cube	Length of one edge cubed	a^3
	Cylinder	Multiply base radius squared by π by height	$br^2\pi h$
	Pyramid	Multiply base length by base width by height and divide by 3	$lwh/3$
	Sphere	Multiply radius cubed by π by 4 and divide by 3	$4\pi r^3/3$
	Cube	Length of one edge cubed	a^3

THE AXIOMS OF EUCLIDEAN PLANE GEOMETRY

For well over two thousand years, people had believed that only one geometry was possible, and they had accepted the idea that this geometry described reality. One of the greatest Greek achievements was setting up rules for plane geometry. This system consisted of a collection of undefined terms like point and line, and five axioms from which all other properties could be deduced by a formal process of logic. Four of the axioms were so self-evident that it would be unthinkable to call any system a geometry unless it satisfied them:

1. A straight line may be drawn between any two points.

2. Any terminated straight line may be extended indefinitely.

3. A circle may be drawn with any given point as center and any given radius.

4. All right angles are equal.

5. If two straight lines in a plane are met by another line, and if the sum of the internal angles on one side is less than two right angles, then the straight lines will meet if extended sufficiently on the side on which the sum of the angles is less than two right angles.

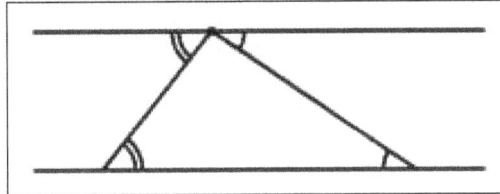

Once we can draw a unique line through one vertex of a triangle not meeting the line containing the opposite side, we can use alternate interior angles to see that the sum of the angles of a triangle is the same as a straight angle, 180 degrees.

Because this axiom was much more complicated than the previous axioms, it seemed more like a theorem than a self-evident proposition. Since all attempts to deduce it from the first four axioms had failed, Euclid simply included it as an axiom because he knew he needed it. For example, some axiom like this one was necessary for proving one of Euclid's most famous theorems, that the sum of the angles of a triangle is 180 degrees. Mathematicians found alternate forms of the axiom that were easier to state, for example:

For any given point not on a given line, there is exactly one line through the point that does not meet the given line.

This form of the fifth axiom became known as the parallel postulate. Although it was simpler to understand than Euclid's original formulation, it was no easier to deduce from the earlier axioms. The attempt to deduce the fifth axiom remained a great challenge right up to the nineteenth century, when it was proved that the fifth axiom did not follow from the first four.

The great advantage of expressing geometry as an axiomatic system was that it no longer was necessary to memorize long lists of independent facts about the nature of the universe--one only had to know a small set of axioms, and by applying to them the rules of inference, one could reconstruct the entire collection of geometric truths.

There was little doubt that the Greeks were attempting to describe a real world when they formulated their geometry, even though it might have been an ideal sort of world, realized only abstractly "in the mind of God." Many mathematicians, now as well as in the distant past, believe that the complete structure of mathematics is something that exists in itself and that it is only gradually discovered by human beings laboring to uncover its mysteries. Even though the framers of the early axiom systems would refer to point and line as undefined terms, they fairly clearly thought of them as real objects, and they thought that the system they were developing was a progressively more and more elaborate and accurate description of the real world. The progress of algebra, on the other hand, was not quite so settled, and people accepted changes in viewpoint there more readily than in the very traditional field of geometry.

EUCLID'S POSTULATES

Any statement that is assumed to be true on the basis of reasoning or discussion is a postulate or axiom. The postulates stated by Euclid are the foundation of Geometry and are rather simple observations in nature. 'Euclid' was a Greek mathematician regarded as the 'Father of Modern Geometry.

Postulate – I

A straight line segment can be formed by joining any two points in space.

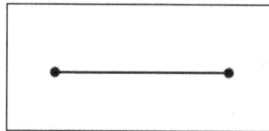

In Geometry, a line segment is a part of a line that is bounded by 2 distinct points on either end. It consists of a series of points bounded by the two endpoints. Thus a line segment is measurable as the distance between the two endpoints. A line segment is named after the two endpoints with an overbar on them.

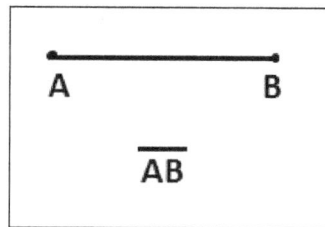

Euclid Geometry Postulate – II

Any straight line can be extended indefinitely on both sides. Unlike a line segment, a line is not bounded by any endpoint and so can be extended indefinitely in either direction. A line is uniquely defined as passing through two points which are used to name it.

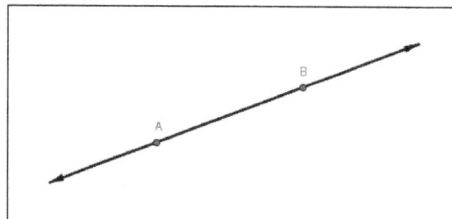

Postulate – III

A circle can be drawn with any centre and any radius. For any line segment, a circle can be drawn with its centre at one endpoint and the radius of the circle as the length of the line segment. Consider a line segment bounded by two points. If one of these points is taken as the centre of a circle and the radius of the circle is taken as equal to the length of the segment, a circle can be drawn with its diameter twice than the length of the line segment.

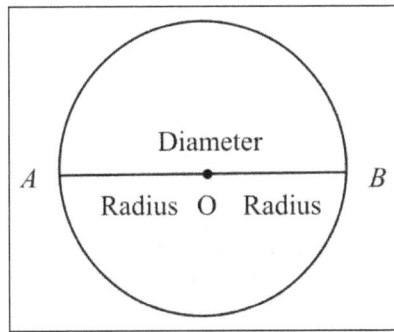

In the above example, the line segment AO serves as the radius of a circle with centre at point O and a diameter equal to AB where l(AB) =2l(AO).

Postulate – IV

All right angles are congruent or equal to one another. A right angle is an angle measuring 90 degrees. So, irrespective of the length of a right angle or its orientation all right angles are identical in form and coincide exactly when placed one on top of the other.

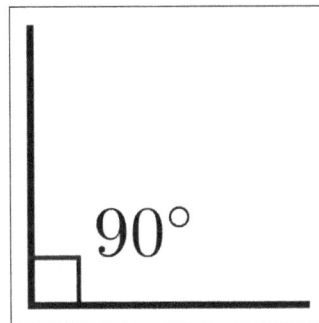

A right angle.

Postulate – V

Two lines are parallel to each other if they intersect the third line and the interior angle between them is 180 degrees.

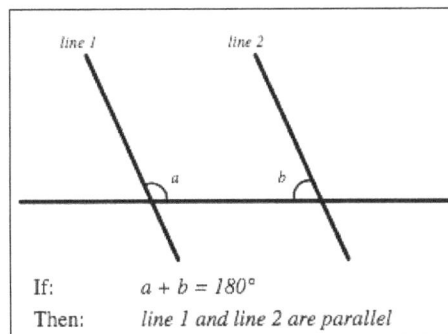

If: $a + b = 180°$
Then: line 1 and line 2 are parallel

Parallel lines' are a set of 2 or more lines that never cross or intersect each other at any point in space if they are extended indefinitely. As you can see in the above image, line 1 and line 2 are parallel if and only if the sum of angles 'a' and 'b' they make with the transversal is 180 degrees.

Permissions

Index

www.ingramcontent.com/pod-product-compliance
Lightning Source LLC
Chambersburg PA
CBHW082051190326
41458CB00010B/3504